ELECTRONICALLY CONTROLLED PROPORTIONAL VALVES

MECHANICAL ENGINEERING

A Series of Textbooks and Reference Books

EDITORS

L. L. FAULKNER

*Department of Mechanical Engineering
The Ohio State University
Columbus, Ohio*

S. B. MENKES

*Department of Mechanical Engineering
The City College of the
City University of New York
New York, New York*

OTHER VOLUMES IN PREPARATION

ELECTRONICALLY CONTROLLED PROPORTIONAL VALVES

Selection and Application

MICHAEL J. TONYAN

Rexroth Worldwide Hydraulics
The Rexroth Corporation
Industrial Hydraulics Division
Bethlehem, Pennsylvania

Tobi Goldoftas, Editor
Hydraulics & Pneumatics

MARCEL DEKKER, INC.

New York • Basel

Marcel Dekker, Inc.
270 Madison Avenue, New York, New York 10016

ISBN: 0-8247-7431-0   RA 00312

Current printing (last digit):
10 9 8 7 6 5 4 3 2 1

Printed in the United States of America

Electronic controls are not new to the hydraulic technology. Indeed, electronically controlled servovalves were used in military equipment as early as the 1940s. Because servovalves were noted for their accuracy, quick response, and remote controllability, they soon became established in the aircraft industry. However, their industrial application remained limited by cost, design complexity, and maintenance problems. Above all, servovalves were extremely sensitive to contamination, and, for that reason, troublesome.

When electronic proportional valves emerged in the early 1970s, they met the needs of many industrial applications. This work was written as an educational tool for the new technology.

Electronically Controlled Proportional Valves covers the specifics of the operation of electronic, proportionally controlled hydraulic components. It also includes the necessary wiring diagrams and setup procedures. For these reasons, this book is an excellent reference for those involved in the maintenance and troubleshooting of in-plant installations which today incorporate electronically controlled hydraulics.

Likewise, because computerized control of hydraulic systems is clearly the way of the future, the timeliness of this text will help prepare and train maintenance personnel and technicians for the future.

Readers involved in the design of new equipment, and those who want to update and modernize existing systems, will find specific application details which must be considered for proper operation of the process being controlled.

Design considerations apply Newton's second law of motion to a hydraulic system to predict actual pressure and flow requirements during acceleration and deceleration. Formulas are also developed with which the designer can determine a system's natural frequency, to assure system operating stability. Finally, application examples consider the various effects resistive and overrunning loads have on the operation of a proportional valve.

This text was especially designed for the reader who has a basic understanding of hydraulic principles and D.C. electrical theory. Note that a strong electronic background is not required to understand these materials. The reader will learn thoroughly how proportional valves work and will develop an understanding of how electronics interface with the controlled functioning of a valve. Many cross-sectional drawings are featured together with a wealth of wiring diagrams, and many hydraulic circuit diagrams, all designed to enhance the subject matter and help the reader master it.

I earnestly believe that *Electronically Controlled Proportional Valves* will help the reader gain greater insights and knowledge of the remarkable and blossoming proportional technology.

Michael J. Tonyan
Bethlehem, Pennsylvania
June, 1985

Time and tide waits for no man. This well-known comment bears heavily on the content of this document.

The hydraulic circuitry of decades past has been rightfully accused of leading to the demise of certain fluid power transmission systems.

Instant gratification has not been the sole tenet of our current generation. Designers in the past can be accused of proliferating "quick and dirty" circuits because they did work. And we add, almost good enough! The same designers of fluid power equipment in the last several decades have leaned heavily on brute strength with marginal control systems to contain the massive power available through the use of a fluid to transmit energy in a predetermined pattern.

Most controls were subject to change associated with thermal factors and machine harmonics were given minimal attention.

Fortunately we have recognized the errors of judgement for what they are. We are relearning and using the laws associated with the acceleration and deceleration of a known mass. We are accepting the aid and assistance of electronic devices to control the action of our pressure level, directional and flow control mechanisms used in hydraulic circuitry.

The proportional hydraulics and associated circuitry documented and explained in this manuscript provide a timely technical update for the practicing engineer and a necessary text for the newly trained technician and/or engineer.

Use of a direct current solenoid or torque motor to interface between signal source and power delivery has turned the tide in a timely manner toward greater use of fluid power as a preferred energy transfer agent.

Side benefits of the devices described herein include smoother, longer lasting machines, improved reliability, and unquestioned repetitive accuracy.

Incidence of hose, tube and pipe breakage is dramatically reduced and service requirements are minimized.

Service functions, if and when needed, are usually fast with minimal machine down time.

This text is a timely bridge into a new era of expanded fluid power systems usage and offers a versatile power transmission method for automatic machinery and associated robotic devices.

J. J. Pippenger
Educational Coordinator
Fluid Power Educational Foundation

CONTENTS

ELECTRONICALLY CONTROLLED PROPORTIONAL VALVES

Chapter 1

PROPORTIONAL VALVES

It is essential that anyone connected with hydraulics understand the differences that exist between servovalves and proportional valves. The two product lines have clearly different characteristics and capabilities and offer diverse advantages. Table 1.1 compares some of the major characteristics of proportional valves and electrohydraulic servovalves.

Table 1.1 Comparison of characteristics of
proportional valves and servovalves

	PROPORTIONAL VALVES	SERVO VALVES
Type of loop	Open	Closed
Feedback	No	Yes
Accuracy	Moderate error factor $\cong 3\%$	Extremely high error factor $< 1\%$
Response	Low: <10 Hz	Very high: 60-400 Hz
Cost	Moderate	High
Need for auxiliary electronic equipment	Moderate	Substantial
Sensitivity to contamination	Tolerant	Highly

Types of Solenoids

There are two types of proportional solenoids. One is stroke-controlled, see page 30, the other is force-controlled. Force controlled solenoids consist of modified DC solenoids which provide linear adjustable forces by altering the current signal to the solenoid, Figure 1.1.

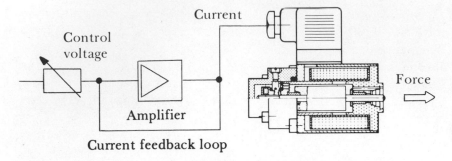

Fig. 1.1 By adjusting potentiometer from 0 to 9 volts, linear output force can be set to maximum force of 14 lb.

Proportional force solenoids are wet pin DC units which tend to resemble conventional DC solenoids, but have a modified internal construction which optimizes the linearity of the solenoid. When a *conventional* DC solenoid is energized, the plunger travels its full stroke, generating a constant output force. A force controlled solenoid operates on the principle that solenoid force output is *linear* with respect to current input. This linear relationship of force-output to current-input works effectively over strokes of about 0.06 in., (1.5mm).

The easiest way to understand a force-controlled solenoid is by studying the force travel curve, Figure 1.2.

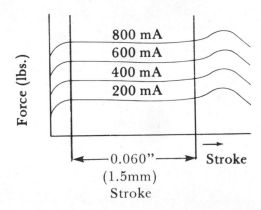

Fig. 1.2 When current to solenoid of proportional valve is held constant, solenoid force also remains constant over stroke length to 0.06 in (1.5 mm).

Since a given amount of current creates a given force, the force-travel curve exhibits this linear relationship at various current levels. When current to the solenoid is held constant, the solenoid force will also remain constant over a stroke output of about 0.06 in., (1.5mm). For example, when a 200-mA signal is raised to 400 mA, the force increases, but remains *constant* over the same stroke output 0.06 in., (1.5mm).

Maximum force output for force controlled solenoids falls in the range from 12 to 14 lb. Since adjustable forces can be achieved over a small stroke, installation dimensions for the solenoid are relatively small. Therefore, the solenoid can generate forces required to pilot-operate proportional pressure control valves, directional control valves, and some variable displacement pump controls. Also, because this is a wet pin type solenoid, it has a removable screw to bleed any trapped air.

PROPORTIONAL PRESSURE RELIEF VALVES

Proportional pressure relief valves perform essentially the same function as manually-adjusted, pilot-operated valves, Figure 1.3 *(a)*. The major difference between the two is that the spring adjustment assembly in the pilot head is replaced with a force solenoid.

Fig. 1.3. (a) Proportional pressure relief valve .

The maximum force of 12 to 14 lb exerted by force solenoids is large enough to enable adjustable forces to hold a small, direct-operated relief valve poppet in a seated position.

When the valve solenoid is energized, the solenoid exerts a direct force on the pilot poppet, Figure 1.3 *(b)*. Pilot pressure fluid flows internally through passage *C*, orifice *D* on the nose of the pilot poppet,

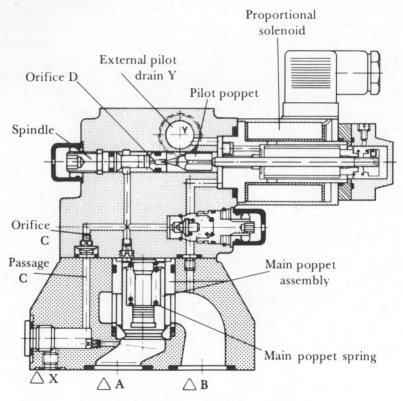

Fig. 1.3 (b) Proportional pressure relief valve cross section.

and to the top side of the main poppet. As long as fluid pressure on the bottom side does not exceed the force to the top side of the proportional solenoid, the main poppet will remain closed because the main poppet has the same area on top and bottom, plus the force exerted by a light spring. Since the forces above and below the main poppet are equal, the downward force exerted by the light spring keeps the poppet seated.

When system pressure exceeds the setting of the proportional solenoid, the pilot poppet opens, allowing pilot fluid to flow from passage C through port Y to tank, Figure 1.3 *(b)*. Pressure decay above the main poppet is felt by orifice C. Simultaneously, the main poppet opens, allowing oil to flow from pressure port A to reservoir port B.

The small, lightweight poppet requires a minimal stroke to open, which means that the valve reacts very quickly. The sleeve has three radial, symmetrically-spaced drilled holes to provide divergent flow characteristics when the valve opens. This results in substantially quieter operation.

Unlike a conventional, pilot-operated relief valve, (in which the maximum pressure rating of the valve is effectively set by the force of the spring in the pilot head), the maximum pressure rating for a proportional pressure relief valve is established by the seat area of the pilot poppet. Since the solenoid provided an adjustable force of 14 lb, a *larger* seat would *lower* the maximum adjustable pressure rating of the valve. Because the resultant force would be determined by a larger area on the nose of the pilot poppet, less pressure would be needed to push it open.

Conversely, a *smaller* seat will result in a *higher* pressure rating of the valve because the resultant force is created by a smaller area on the nose of the pilot poppet: it takes a greater pressure to open the pilot poppet. The seat area is normally fixed at the factory when the maximum pressure rating of the valve is set.

The sensitivity of the proportional solenoid requires that the pilot head be drained externally *directly* back to tank through port Y. If the pilot head were drained internally, back-pressure could cause erratic valve operation.

The electronic amplifier makes it now possible to increase or decrease gradually the pressure setting of the valve. Also, the solenoid force can be adjusted quickly and frequently during machine operation. Adjusting time (time needed to go from one valve setting to another in response to the signal received from the amplifier) is in the range from 50 to 159 ms, depending on valve size.

In case of electrical power failure, the solenoid force drops immediately, allowing oil to flow from port A to port B. To protect the system from high, unexpected solenoid forces, (as may be caused by electronic failure or high current peaks), a maximum pressure relief valve can be built into the valve pilot head. This can be accomplished with a standard spring-and-poppet assembly which can be adjusted mechanically just slightly above maximum desired system pressure.

Remember that a relief valve is often needed for very small flow rates. By removing the pilot head and using it as an electronically adjustable, direct-operated relief valve, the pilot head alone can handle flows to 0.5 gpm.

PROPORTIONAL PRESSURE REDUCING VALVES

Proportional, pilot-operated pressure reducing valves, Figure 1.4 *(a)* are similar to proportional pilot operated pressure relief valves, in that they are electronically adjustable and incorporate a force solenoid. The main valve assembly is the same as in a manually adjusted, pilot operated, pressure reducing valve, with the pilot head the same as in a proportional relief valve. The maximum pressure rating of the valve is, therefore, determined by the seat of the pilot poppet as in the relief valve.

When the valve receives an input signal, the proportional solenoid acts directly on the pilot poppet. As long as the force exerted by the solenoid holds the pilot poppet closed, the pilot oil remains in a static condition. Pressure fluid flows through passage C, Figure, 1.5,

Fig. 1.4 (a) Proportional pressure reducing valve.

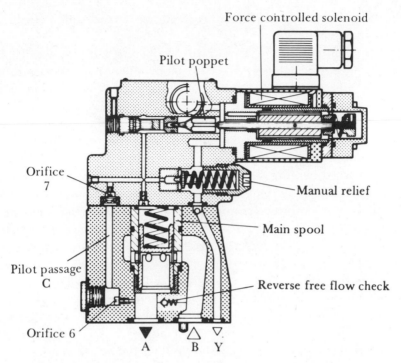

Force controlled solenoid

Pilot poppet

Orifice 7

Manual relief

Main spool

Pilot passage C

Reverse free flow check

Orifice 6

A　　B　Y

Fig. 1.4 (b) Proportional pressure reducing valve cross-section.

and acts above and below the main spool. Since the areas of the main spool are equal, a force balance is created. Because the main spool is balanced hydraulically, a light spring force can hold the main spool open: oil flows unrestricted from the primary port *B* to secondary port *A*. When the force acting in the secondary port exceeds the force exerted by the proportional solenoid, the pilot poppet opens, draining oil from the top of the main spool to tank, via port *Y*.

This creates a pressure drop in orifices 6 and 7 to unbalance the main spool, Figure 1.4 *(b)*. The main spool now moves up, reducing the flow area from port *B* to port *A*, (through the radial holes in the main spool and sleeve) and creating a secondary reduced pressure in port *A*. The main spool then modulates to maintain pressure in port *A* at the setting of the proportional solenoid.

The electronic amplifier can adjust the solenoid force quickly and/or frequently during machine operation. Adjustment time (the time required to go from one valve setting to the next consistent with the signal received from the amplifier) ranges from 100 to 300 ms, depending on valve size. Solenoid force can also be increased or decreased gradually, resulting in a gradual increase or decrease in pressure, as needed.

For maximum safety, a manually adjusted relief valve can be installed in the pilot head. As with the proportional pressure relief valve, the pilot head should be drained externally, directly back to tank.

An electronically adjustable proportional pressure reducing valve equipped with a pressure compensated flow controller in the pilot head and overload protection can maintain precise pressure settings at its highest flow rating. Since the main spool in the pressure reducing valve acts like a regulating orifice (at high flows), flow *downstream* of the orifice is likely to be more turbulent. To make the valve more precise at high flow, pilot oil is taken from the primary port rather than the secondary, preventing turbulent flow from influencing the valve setting.

As was the case with the proportional pressure reducing valve, as long as the solenoid force keeps the pilot poppet *closed,* the valve will remain *open*, allowing oil to flow from primary port B through the main spool, to secondary port A, Figure 1.5. This open position is maintained by supplying oil from primary port B, through passage C, through the pressure compensated flow controller to the top (spring loaded side) of the main spool. Desired system pressure is thus maintained on top of the poppet, creating an unrestrictd flow path from port A to port B because the main spool and sleeve are aligned through their respective radial holes.

When the force exerted by pilot pressure exceeds the force exerted by the proportional solenoid valve, the pilot poppet opens. The pressure compensated flow controller establishes a constant pilot flow and oil drains through port Y to tank, creating regulated pressure above the poppet.

When pressure in secondary port A exceeds the combined forces exerted by the spring and pilot pressure, the main spool moves upward, creating a secondary reduced pressure, and modulates, as needed. If pressure in the secondary port rises too high during a static condition, the overload protection will open, allowing oil to return to the pilot head, thus preventing pressure build up because of leakage.

Both types of proportional pressure reducing valves are available with reverse free flow checks.

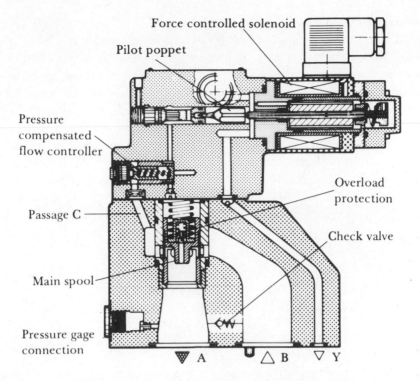

Force controlled solenoid

Pilot poppet

Pressure
compensated
flow controller

Overload
protection

Passage C

Check valve

Main spool

Pressure gage
connection

▽ A   △ B   ▽ Y

Fig. 1.5 Proportional pressure reducing valve with compensated flow controller.

PROPORTIONAL DIRECTIONAL VALVES

Proportional, 4-port directional control valves are the most versatile of all. The valve looks like a conventional directional valve, but its spool configuration has been modified to provide precise metering in the inlet and outlet sections of the valve. Not only can this valve meter oil in both directions, but when used properly, pressure drops on both sides of the valve remain relatively equal. This allows for good controlability of cylinders and motors.

Acceleration, deceleration and counterbalancing can often be achieved with only one proportional directional valve, when interfaced with appropriate electronic controls. With conventional valving, such controlability is sometimes not possible even when as many as seven valves are used.

Proportional Valve Spool Types

The radial clearance between the proportional valve bore and spool is held to about 3 to 4 micrometres (0.00012 to 0.00016 in.). Although spool valves experience some internal leakage, these close tolerances hold leakage to a minimum and allow for spool overlap to be held at a minimum. Total spool overlap is exactly 11%.

Spool configurations for 4-port proportional valves, Figure 1.6 are basically simple in construction and easy to specify. As discussed earlier, the spools are designed to provide metering in both directions of flow.

Fig. 1.6 Spool configuration for closed-center, 4-port proportional valve with 1:1 area ratio. V-shaped notches meter flow.

Figure 1.7 illustrates a closed center spool configuration. The triangular metering notches in the spool lands are sometimes called control grooves. Each land has eight control grooves cut symmetrically around the periphery and thus providing equal areas in both directions. The number of control grooves will vary depending on the application.

When the spool is shifted in either direction, the control grooves never fully clear the ports, thus always maintaining a metering function. As in conventional directional valves, the spool first moves through a deadband, then opens fully, to virtually eliminate metering.

Although all proportional spools overlap exactly 11%, compensation in the amplifier reduces this value to a minimum. The spool is normally used to control linear and rotary actuators. Since it is an equal area spool, pressure drops from ports *P* to *B* or from ports *P* to *A* remain fairly equal, providing good controlability.

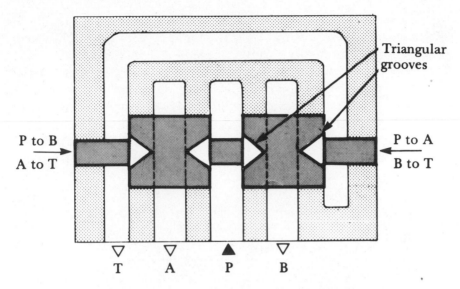

Fig. 1.7 V-shaped grooves at both ends of spool provide metering
when spool is shifted in each direction.

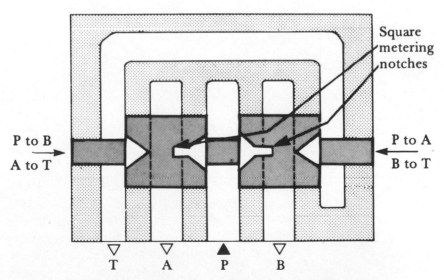

Fig. 1.8 Square metering notches provide restricted center
position from ports P to A and B, with port T blocked.

Figure 1.8 illustrates a restricted center spool which provides a restricted flow path from ports *P* to *A* and *B*, with port *T* blocked. The restricted center position is achieved with square metering notches on both spool lands, allowing about 3% metered oil of the full flow rating of the spool. The spool is normally used to control hydraulic motors to provide the necessary makeup oil in center position. Makeup fluid may be needed because of motor leakage or any suction that may be caused when the motor stops suddenly. Other than the modified center position, spool construction and operation is the same as that shown in Figure 1.7.

Figure 1.9 shows a restricted center spool, which provides a restricted flow path from ports *A* and *B* to *T* with port *P* blocked. While in center position, the square metering notches again provide

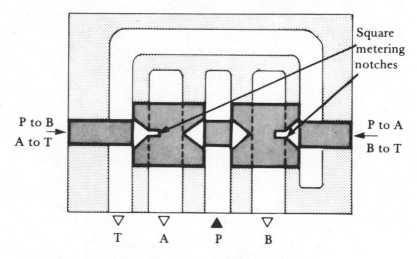

Fig. 1.9 *Square metering notches provide restricted center position from ports A and B to T, with port P blocked.*

3% of metered flow. The spool is normally used with single rod cylinders with an area ratio close to 1:1. Since all spool valves experience some leakage, oil leaking from port *P*, when the valve is centered, can drain from ports *A* and *B* directly to tank. This arrangement eliminates the risk of inadvertent cylinder extension and pressure intensification. When the spool is used for overhung loads, some type of counterbalance or pilot operated check valve must also be used, see Chapter 4.

A regenerative, closed center spool is shown in Figure 1.10. Because the right outside land , C, has no control grooves, it blocks ports B to T when the spool is shifted to the left. A regenerative

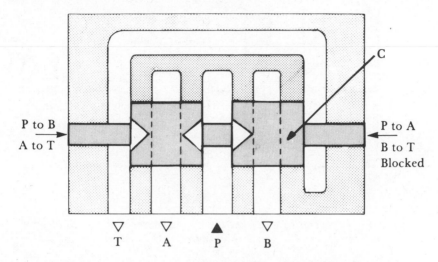

Fig. 1.10 *Regenerative, closed center spool configuration has no control grooves in face C, blocking ports B to T when spool shifts to left.*

spool with a restricted center, Figure 1.11, allows oil to bleed from ports A and B to port T, while port P is blocked. The right spool land is extended with a square metering notch to meter oil from port B to port T when the spool is centered. When the spool is shifted to the left, flow from port B to port T is blocked.

Spools with 2:1 Area Ratios

When selecting a proportional spool to actuate cylinders which have *pistons* with area ratios close to 2:1, the system designer must consider several factors.

In a cylinder with a 2:1 area ratio, the head end delivers half the amount of oil that enters the cap end. Conversely, the cap end delivers twice the volume of oil that flows into the head end. If an *equal* area closed- or restricted-center spool is used to control a cylinder with a 2:1 area ratio, the pressure drop across the valve is likely to be *unequal* in *both* directions. This condition can lead to serious cylinder control problems.

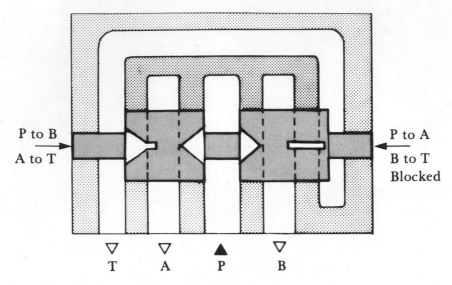

P to B
A to T

P to A
B to T
Blocked

▽ T   ▽ A   ▲ P   ▽ B

*Fig. 1.11 Regenerative spool with restricted center allows oil to bleed
from ports A and B to T, with port P blocked.*

To avoid this situation, a closed center or restricted center spool with a reduced area can be specified to supply half the volume on one side of the spool, as compared to the other, see Table 1.2.

Figure 1.12 shows how this reduced flow area is achieved. By eliminating half of the control grooves on land A, the flow area becomes half of that of land B. This arrangement keeps the total pressure drop across the valve fairly equal, thus maintaining good controlability of cylinders which have pistons with area ratios close to or equal to 2:1.

*Fig. 1.12 Closed center spool configuration with
reduced 2:1 flow area ratio.*

Table 1.2 Flow paths for 3 position proportional spools

Symbol	Type	Flow paths	Application
A B P T	CLOSED CENTER	$P \rightarrow A = Q, B \rightarrow T = Q$ $P \rightarrow B = Q, A \rightarrow T = Q$	Motor/cylinder spool (with cylinder area ratio close to 1:1)
	CLOSED CENTER	$P \rightarrow A = Q, B \rightarrow T = Q/2$ $P \rightarrow B = Q/2, A \rightarrow T = Q$	Cylinder spool (with cylinder area ratio close to 2:1)
	CLOSED CENTER	$P \rightarrow A = Q/2, B \rightarrow T = Q$ $P \rightarrow B = Q, A \rightarrow T = Q/2$	Cylinder spool (with cylinder area ratio close to 2:1)
	CLOSED CENTER	$P \rightarrow A = Q, B \rightarrow T$ Blocked $P \rightarrow B = Q, A \rightarrow T = Q$	Regenerative spool
	RESTRICTED CENTER $P \rightarrow A$ and B T Blocked	$P \rightarrow A = Q, B \rightarrow T = Q$ $P \rightarrow B = Q, A \rightarrow T = Q$	Motor spool
	RESTRICTED CENTER A and $B \rightarrow T$ P Blocked	$P \rightarrow A = Q, B \rightarrow T = Q$ $P \rightarrow B = Q, A \rightarrow T = Q$	Cylinder spool (with cylinder area ratio close to 1:1)
	RESTRICTED CENTER A and $B \rightarrow T$ P Blocked	$P \rightarrow A = Q, B \rightarrow T = Q/2$ $P \rightarrow B = Q/2, A \rightarrow T = Q$	Cylinder spool (with cylinder area ratio close to 2:1)
	RESTRICTED CENTER A and $B \rightarrow T$ P Blocked	$P \rightarrow A = Q/2, B \rightarrow T = Q$ $P \rightarrow B = Q, A \rightarrow T = Q/2$	Cylinder spool (with cylinder area ratio close to 2:1)
	RESTRICTED CENTER A and $B \rightarrow T$ P Blocked	$P \rightarrow A = Q, B \rightarrow T$ Blocked $P \rightarrow B = Q, A \rightarrow T = Q$	Regenerative spool

Pressure drop calculations for cylinders with 2:1 piston area ratios are shown on page 71 along with cavitation effects and braking pressure when a spool with a 1:1 area ratio is used with a cylinder which has a piston with a 2:1 area ratio.

PROPORTIONAL 3-PORT PRESSURE CONTROL VALVES

Before discussing pilot-operated, proportional directional control valves, Figure 1.13, let us first examine the pilot head for this valve.

To understand the total pilot-operated proportional directional control valve, the designer must first fully understand the pilot head which, essentially, consists of two 3-port proportional pressure reducing valves. A simplified diagram of a 3-port proportional pressure reducing valve is shown in Figure 1.14.

When the valve solenoid is energized, a corresponding force shifts the spool to the right. This action connects port P to port

Fig. 1.13 Three-port, proportional pressure control valve.

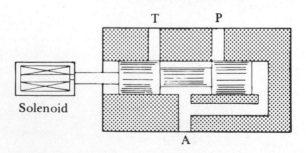

Fig. 1.14 Simplified diagram of 3-port proportional pressure reducing valve.

A, enabling the oil to perform its intended function. At the same time, fluid pressure rises in port *A* and is fed to the opposite end of the spool exerting a counterforce which acts against the force exerted by the solenoid. When fluid pressure in port *A* is high enough to exert a force *equal* to the force exerted by the solenoid, the spool shifts to its center or neutral, no-flow position.

If pressure in port *A* exerts a force that *exceeds* the solenoid force, the spool shifts to the left, allowing oil to flow from port *A* to tank *T* until equilibrium is restored between the two forces, again, allowing the valve to return to center position. This basically describes the action of the pilot head. The only difference is that the valve incorporates two proportional force solenoids, and consists of a special 3-piece spool arrangement.

As Figure 1.15 shows, the solenoid force acts directly on this 3-piece assembly, which consists of a control spool with a free moving sensing piston in each end of the spool. Both sensing pistons are free to move in each end of the control spool.

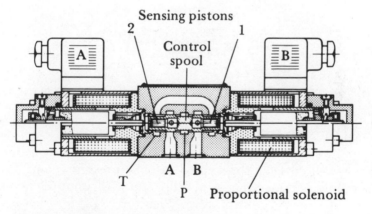

Fig. 1.15 Cross section of typical pilot operated proportional directional control valve.

If solenoid *B* is energized, the force exerted by the solenoid pushes directly against the sensing piston -1, which shifts the control spool to the left. This spool shift allows oil to flow from port *P* to port *A* and cause fluid pressure to rise in port *A*. At the same time, two radial holes in the control spool allow pressure fluid in port A to flow through the drilled hole furthest to the left, thus acting on sensing piston-2. Since sensing piston-2 is free to move in the end of the control spool, pressure fluid pushes sensing piston-2 out against solenoid A. A photograph of the spool and sensing pistons is shown in Figure 1.16.

The pressure forces between sensing piston-2 and the control spool work against the force exerted by solenoid B. When pressure in port A rises (to where it exerts a force equal to that of solenoid B), the control spool shifts to the right, blocking the connection from port P to port A, while holding pressure in port A constant. If the force exerted by solenoid B is reduced, pressure fluid in port A shifts the spool even farther to the right. Oil can now drain from port A to port T until pressure is reduced to the level where it again corresponds to the force exerted by the solenoid. If a signal is provided to solenoid A, the process reverses with port P being connected to port B.

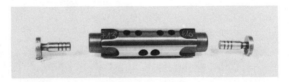

Fig. 1.16 Control spool and pistons used in valve illustrated in Fig. 1.15.

PILOT OPERATED PROPORTIONAL DIRECTIONAL VALVES

Having discussed 3-port proportional pressure control valves, we can now consider the overall operation of pilot-operated, proportional directional control valves, Figure 1.17 *(a)*.

Electronically adjustable, pilot operated proportional directional valves can be piloted internally or externally through port X, Figure 1.17 *(b)*. Pilot pressure requirements must be evaluated carefully. To make sure that the main spool will open fully under *all* operating conditions, a minimum pilot pressure of 435 psi is required at the pilot valve inlet X. Also, if the valve is piloted *internally* and the system operates at pressures above 1450 psi, a sandwich mounted pressure reducing valve *must* be installed between the main valve and

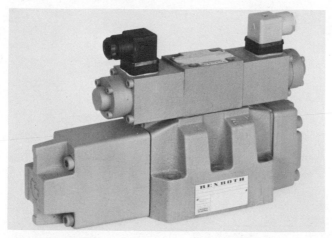

Fig. 1.17 (a) Pilot operated proportional directional valve.

the pilot section. The sandwich element is needed to protect the pilot section as its maximum operating pressure at port P is limited to 1450 psi.

When both solenoids are de-energized, Figure 1.17 *(b)*, the main spool (closed center with a 1:1 area ratio) is held in its center position by a compressed push-pull spring. A pressure signal in pilot chamber C shifts the spool to the left, compressing the spring. Likewise, a pressure signal in chamber D shifts the spool to the right, compressing the spring against the valve housing or pulling on the spring assembly.

The spool remains spring-centered until pilot pressure in one of the end caps can generate enough force to shift the spool to a metering position. Maximum pilot pressure needed in either chamber to move the spool is 365 psi. Then, if solenoid B receives its maximum signal, the solenoid develops 14 lb of force and moves the control spool to the right allowing a 365 psi pressure signal to develop in chamber C. Simultaneously, the main spool is moving toward the left a distance *proportional* to the pressure signal developed by the force solenoid.

Not until the force of the solenoid equals that exerted in chamber C, does the pilot hold constant pressure in the pilot chamber. The main spool then maintains a set position which, in this case, would cause the spool to travel its *full* stroke. Since the spool moves in proportion to the input signal received by solenoid B, the control

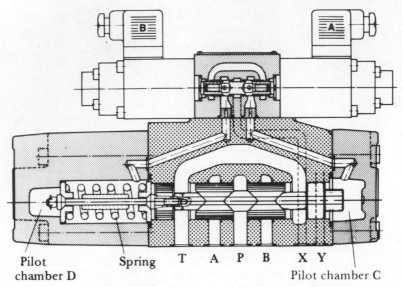

Pilot
chamber D   Spring   T   A   P   B   X   Y
                                        Pilot chamber C

*Fig. 1.17 (b) Cross-section through pilot operated proportional directional
control valve.*

grooves open progressively, allowing gradually increasing flow from
port P to port A, and from port B to port T. If a signal is received at
solenoid A, the process is reversed and the main spool shifts to the
right, allowing oil to flow from port P to port B and from port A to
port T.

Adjustable forces can be obtained by controlling the signal
level supplied, making it possible to shift the valve spool to various
preselected positions. Because the metering control grooves are
triangular, each time a particular spool position is reached, a corresponding orifice is created. Various actuator speeds can thus be
preselected by presetting the signal level to the valve.

With the help of the electronic amplifier, a time-controlled movement of the spool can provide smooth load starting and stopping.
Thus, with the pilot operated proportional directional valve just
discussed, it would be easy to accelerate a load to a predetermined
constant velocity, then decelerate it to a stop within a time-controlled
stroke of the spool and precise metering.

Assume that solenoid B, Figure 1.17 *(b)* requires a 100% signal
to determine the set point. When solenoid B receives a 100% signal,
the signal can be increased electronically from zero to 100% at a

given time rate. This means that the force of the solenoid is increasing gradually and that a pressure signal is developing gradually in pilot chamber C. Simultaneously, the spool is moving proportionally from its center position to its set point, progressively opening port P to port A and port B to port T, while accelerating the load to a constant velocity.

As the load decelerates to a stop, solenoid B is de-energized, reducing the signal level from 100% to zero at a preset time rate and moving the spool back proportionally from its set point to its center position while stopping the load smoothly.

Clearly, as the signal increases from zero to 100% or decreases from 100% to zero, the spool responds accordingly to the increasing or decreasing time-controlled signal. Thus, the rate for which the signal is set determines the rate at which the spool reaches its set point. The amount of signal determines the final set point of the valve or final spool position.

From the above discussion it can be concluded that:

1. the signal level determines final spool position,
2. with the help of an electronic amplifier, a time-controlled signal determines how quickly or slowly the spool will reach its set point, directly proportionally to acceleration or deceleration, and
3. the spool provides metering in both directions.

HYDRAULICALLY OPERATED PROPORTIONAL DIRECTIONAL VALVES

Although many applications require the type of control offered by a proportional directional valve, designers appear to resist incorporating the associated electronics. A direct, pilot-operated proportional directional valve without electronics can be used if the application requires only simple, direct control: its main valve is identical to the electronically-controlled, pilot-operated directional valve. The only difference is in the pilot head which is replaced with a connecting plate, Figures 1.18 (a) and (b).

Now, the electronically controlled proportional valve becomes a hydraulically controlled proportional directional valve which requires external pilot pressure.

The connecting plate connects pilot port A to port Y and pilot port B to port X; a pilot pressure signal at port X shifts the spool to

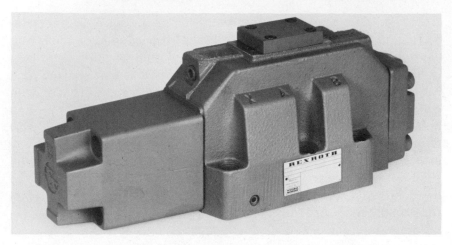

Fig. 1.18 (a) Hydraulically operated proportional directional valve.

the right to provide a flow path from port P to port B and from port A to port T. Pilot pressure at port Y shifts the spool to the left, to provide a flow path from port P to port A and from port B to port T. Spool movement is proportional to a pilot pressure signal ranging from 21 to 365 psi.

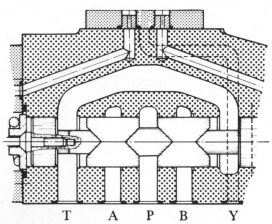

T   A   P   B   Y

Fig. 1.18 (b) Cross-section through pilot head showing connecting plate.

Commonly used pressure controls to generate pilot pressure signals include joysticks, pedals, and pressure reducing valves. Any of these can be mounted remotely from the main valve, thus providing convenient installation at control consoles and test areas.

OPERATING CURVES FOR
PROPORTIONAL DIRECTIONAL VALVES

To provide full control for proportional directional control valves, the inlet and outlet must be metered continually. To achieve good resolution, maximum possible stroke must be used. Various nominal flow ratings are available for each valve size. The various flow ratings are obtained by either increasing or decreasing the size or number of control grooves in the spool.

For each nominal flow rating, an operating curve has been developed to make sure that the intended use of any spool provides for maximum controlability.

Operating curve 1, Figure 1.19 shows a flow-to-input control current percentage relationship for a 26.4 gpm nominal spool. Consider, for example this condition. Assume that a flow of 25 gpm is required and that the valve must shift from a fully closed position to fully open. The curve for 100% control current and a flow of 25 gpm indicates that the pressure drop across the valve is curve 1, or about 150 psi. This would mean a pressure drop of 75 psi from pump port P to actuator port A and another 75 psi pressure drop from actuator port B to tank port T, thus using the full spool stroke.

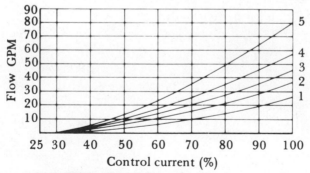

26.4 GPM nonimal flow at 145 psi across valve

1 Δ p = 145 psi drop	4 Δ p = 725 psi drop
2 Δ p = 290 psi drop	5 Δ p = 1450 psi drop
3 Δ p = 435 psi drop	

Fig. 1.19 Family of operating curves shows percentage relationship of flow-to-input control current for 26.4-gpm nominal spool.

If, however, only 10 gpm were required, and the same 26.4 gpm nominal spool were used, curve 1 at 70% control current shows that the valve is already passing 10 gpm at 150 psi pressure. Anything beyond 70% control current does not effectively use full spool stroke. The valve would virtually pass 10 gpm, unmetered, which would, in essence, provide little or no controlability at the end of the curve. A smaller spool would be needed for full control.

The curve clearly demonstrates that the purpose of using proportional control valves is to provide control. To provide full control, there must also be metering, meaning there must be a pressure drop across the valve. Note that the above example does not consider load conditions. A more detailed explanation of these curves is covered later in this text.

SPOOL MOVEMENT WITH STEPPED INPUT SIGNAL

When predicting the maximum cycle rate and load condition of a particular circuit, it is sometimes necessary to consider the physical limiting factors of the valves used. Although the natural frequency* of the system (and therefore the maximum acceleration rate) usually becomes the limiting factor, it is important to become familiar with the response characteristics of the valve.

This is especially true when process controllers or computers are used to control the time relationships between component functions. By being able to predict a dependable valve response characteristic, computerized control can repetitively anticipate the required starting point of a function. The computer or controller can then shift the valve *before* a required function takes place. In this way deadband is eliminated and cycle rates can be improved.

Figures 1.20 *(a)*, *(b)*, and *(c)* show the fastest possible spool shift from one position to another, assuming a stepped (immediate change) input to the electronic amplifier.

Figure 1.20 *(a)* shows a change in stroke command from zero to 100% on the left of the graph and a change in signal from 100% to zero on the right. Figures 1.20 *(b)* and *(c)* are similar. However,

---

*Natural frequency is discussed in detail in Chapter 2.

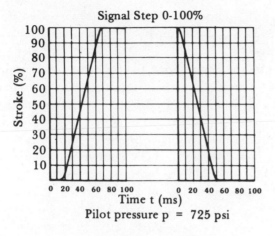

Signal Step 0-100%

Pilot pressure p = 725 psi

(a) Change in command of 100% from 0 to 100% along left y-axis and from 100% to 0 along right y-axis.

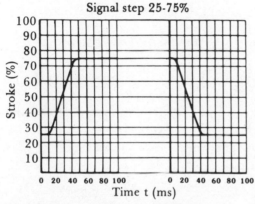

Signal step 25-75%

(b) Change in command: 50% stepped signal from 25 to 75%.

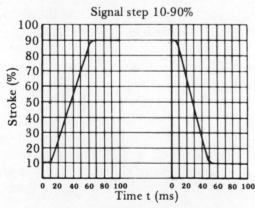

Signal step 10-90%

(c) Change in command: 80% stepped signal from 10% to 90%.

Fig. 1.20 Fastest possible spool shift time from one position to another.

Figure 1.20 *(b)* shows a 50% stepped signal change (25 to 75% along the *y* axis) and Figure 1.20 *(c)* shows an 80% change in command (10 to 90% along the *y* axis).

It is important to note that the above figures represent graphically the valve reaction to a single change in input signal. The *x* axis shows the time required, in milliseconds, for the valve to reach a desired position, assuming an immediate command to shift to that position. This time requirement introduces phase lag in the system which is important when considering a cyclic change in command.

In Figure 1.20 *(a)*, the graph shows that for the valve to shift from zero to 100%, the signal must last 80 milliseconds. Likewise, when the signal is removed, no new command can be introduced for at least 70 milliseconds, if the valve is to close fully. In other words, it takes about 150 milliseconds to complete one cycle. This means:

$$(1 \text{ cycle}/150 \text{ ms}) \times (1000 \text{ ms/sec}) = 6.6 \text{ cps}$$

Thus, the least time the valve can be cycled and achieve 100% response is 6.6 cps. Should the signal change cyclically at a faster rate, the valve can no longer keep up with the command. For instance, if the signal changes at 10 cps, the zero to 100% command to the valve tells the valve to open. The valve, however, may be able to open only 50% before it receives a command to close.

Frequency Response Curves

To explain the frequency response characteristics of the valve, it is first necessary to define the terms used to describe these performance measurements.

Frequency is the number of times an action occurs in a given measure of time. Thus, when frequency is measured in Hz (cps), we are speaking of the number of cycles per second. As used in this text, we refer to the frequency of a command signal in Hz.

Amplitude response is the ratio of output change to input change, and is measured in db (decibels).

Decibel is the $\log_{10}$ of the ratio of output change to input change. Logarithms are used as a convenience to condense numbers. Here are two examples:

Example 1: Assume an input change of 100% and a measured

output change of 50%. What would be the amplitude response?

Amplitude response (db) = $(\log_{10} \times 50\% \times 10)/100\% = -3\text{db}$

The negative sign indicates that the output is decreasing since the $\log_{10}$ of a fractional number is negative.

Example 2: Assume an input change of 100% and an output change of 100%. What will be the amplitude response?

Amplitude response (db) = $(\log 100/100)10 = \text{zero db}$.

Phase lag is the time required for the output to recreate the command of the input. In cyclic occurances, this is normally measured in degrees, Figure 1.21.

(a)

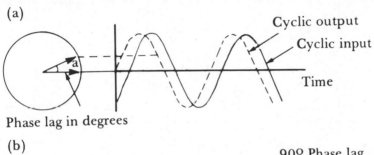

Phase lag in degrees

(b)

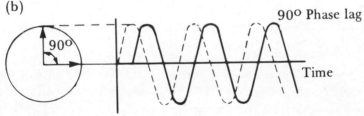

(c)

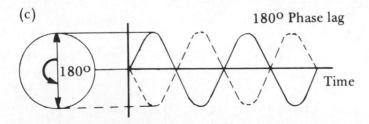

Fig. 1.21 Phase lag is time required for output to recreate input command signal (a). At 90° phase lag (b). At 180° phase lag (c), system becomes unstable.

Note: At 180° of phase lag, Figure 1.21 *(c)*, the system becomes unstable because the output does exactly the opposite of the command; as the command grows larger, the output grows smaller and vice versa.

Figure 1.22 shows a typical frequency response curve for a proportional (or electrohydraulic servo) valve. Curves *(a)* and *(b)* show the relationship between amplitude response and frequency. Curve *(a)* shows a $\pm$ 25% change assuming a 50% input (50% opening port P to port A and port B to port T). Curve *(b)* shows a 50% signal which varies to a high of 100% and a low of 0% cyclically.

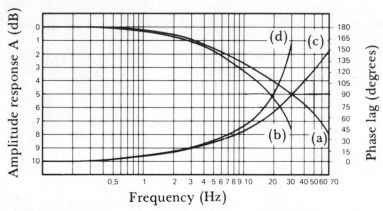

Fig. 1.22 Frequency response curves for proportional or electro-hydraulic servovalve.

For this particular valve, at an amplitude of -3db (50% output), the frequency response is 10 Hz. It is important to note that the industrial standard for accompanying valves has been placed at the -3db level. The curves in Figure 1.22 clearly indicate that as the signals accelerate, the amplitude of spool motion in the wave decreases.

Bottom curves (*c* and *d*) in Figure 1.22 show the relationship between phase lag and frequency. Phase lag as measured in degrees, is shown on the right vertical axis of the graph. As signal frequency increases, the ability of the valve to keep up decreases. In other words, the phase lag becomes greater. In comparing curves *(c)* and *(d)* (variations in the amount of signal change as previously discussed) it takes considerably longer (more degrees of phase lag) for the spool to catch up as the amount of required spool motion increases.

ELECTRONICALLY CONTROLLED DIRECT OPERATED PROPORTIONAL DIRECTIONAL VALVES

Up to now, two types of proportional directional valves were discussed: electronically controlled pilot operated directional valves and hydraulically controlled proportional directional control valves. A third type, used for lower flow ranges, is an electronically controlled, direct operated proportional directional valve, Figures 1.23 *(a)* and *(b)*.

The basic construction of the valve resembles that of the 3-port proportional pressure control valve. Both are equipped with

Fig. 1.23 (a) Electronically controlled, direct-operated proportional directional valve.

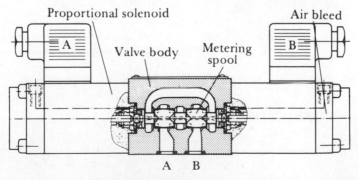

Fig. 1.23 (b) Partial cross-section through body and spool.

force solenoids. In the valve in Figure 1.23 *(b)*, the 3-piece spool assembly is replaced with a proportional spool. In this case, when either solenoid is energized, the force exerted by the solenoid acts *directly* on the spool, and positions the spool against the centering spring forces.

When both solenoids are de-energized, two opposite acting return springs hold the spool in its center position. If solenoid A is energized, the signal shifts the spool to the right a distance proportional to the force of the signal solenoid A received, allowing oil to flow progressively from port P to port B, and from port A to port T. The opposite happens when solenoid B is energized, allowing oil to flow from port P to port A and from port B to port T.

STROKE CONTROLLED SOLENOIDS

A proportional solenoid with a built-in positional transducer (linear variable differential transformer - LVDT) is called a stroke-controlled solenoid, Figure 1.24. These solenoids were developed to

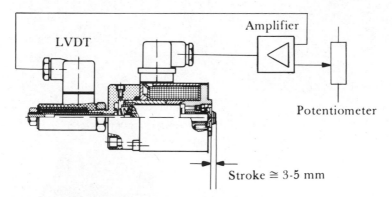

Fig. 1.24 *Proportional solenoid with built-in positional transducer (LVDT) is called a stroke controlled solenoid.*

improve valve performance, which the LVDT makes possible as it provides the electrical feedback which allows the solenoid stroke to be measured more accurately.

When a proportional solenoid first receives an input signal, a certain percentage of error is generated if direct operation of a process is required. Such would be the case with a directly controlled spool stroke of a directional control valve. Because the LVDT is built into the end of the proportional solenoid, it measures the actual spool position, then feeds back a signal to the amplifier.

The amplifier compares electronically the input and feedback (or actual) signals and supplies a corrected signal to the solenoid to compensate for any error. This results in the flow force on the spool or orifice being counteracted, maintaining a very accurate spool position or orifice flow area.

Because every solenoid is subject to some friction, with the use of the LVDT friction is compensated to provide low hysteresis and good repeatability. This is another important characteristic of stroke-controlled solenoids.

DIRECT OPERATED PROPORTIONAL
DIRECTIONAL VALVES WITH FEEDBACK

So far, our discussion about proportional directional valves has been limited to directional valves with force-controlled solenoids. Proportional directional valves that incorporate positional feedback can be operated directly with a high degree of accuracy.

Figure 1.25 *(a)* illustrates a stroke-controlled, direct-operated proportional valve; its spool, Figure 1.25 *(b)*, is held in center (neutral) position by two centering springs. When solenoid *A* is energized, the spool shifts to the right, connecting ports *P* to *B* and port *A* to port *T*. When solenoid *B* is energized, the spool shifts to the left, connecting ports *P* to *A* and port *B* to port *T*. The LVDT being

Fig. 1.25 (a) Direct operated proportional directional valve with feedback.

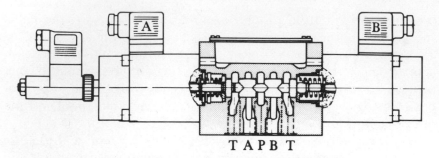

T A P B T

Fig. 1.25 (b) Partial cross-section through body and spool.

linked mechanically to solenoid A can move $\pm$ 0.117 in. for either direction of the spool.

Thus, when either solenoid is energized, the spool shifts a corresponding distance, moving the core of the LVDT, (which, as mentioned, is linked mechanically to solenoid A) out of equilibrium. This, in turn, induces a signal and feeds it back to the amplifier, relaying to it the actual spool position.

The input and feedback (actual) signals are now compared electronically in the amplifier. From these two signal values, a correct signal is generated and fed to the solenoid, giving a definite spool position. If feedback is lost, the spool will return to its center position. This safety feature is built into the amplifier.

The main spool, like that in the pilot operated proportional directional valve, has control grooves cut into it for progressive flow action. Unlike the pilot operated directional valve, there is no need for pilot pressure actuation because the valve is operated directly.

It is also important to mention that although stroke-controlled directional valves are the most accurate of all proportional directional valves, they, however, have some drawbacks, namely their size. The largest available conventional valve has a nominal flow of 16 gpm at a pressure drop across the valve of 150 psi. By comparison, pilot operated, proportional directional valves are available with nominal flows of 137 gpm, with corresponding pressure drops of 150 psi.

Stroke controlled proportional directional valves could be made larger to handle larger flows. However, higher flow forces would be generated at higher flow rates, creating other problems. With the solenoid acting directly on the valve spool, the solenoid would become very large to counteract the flow forces generated. Although the electrical feedback would try to maintain position, the solenoid would eventually run out of force.

Conversely, with pilot-operated, directional control valves, the control area of the main spool is a function of pressure. The resultant pressure forces are much higher, which means that the error generating forces or flow forces are a much smaller percentage.

For example, if a direct operated proportional directional valve provides 70 gpm at full flow through the valve, flow forces might exert a force of 5 lb on the spool. Considering that a direct operated proportional directional valve operates directly on the spool with a force of 14 lb, a substantial amount of flow force would be counteracting the solenoid force. While the electrical feedback is trying to shift the spool back to its proper position, flow forces generated could still become large enough to overcome the force of the solenoid. Percentage wise, the error would become substantial.

For example, a pilot operated proportional directional valve which might be operating at a pilot pressure of 365 psi, acts on the area at each end of the spool. Since force equals pressure times area, the resultant force would be high enough to counteract a flow force of 5 lb, creating a substantially smaller error.

This is the reason why pilot-operated, proportional directional valves can be built without feedback and can handle higher flows. As valve size increases, the effective area on the end of the spool also increases. This increase in area means that the pilot pressure of 435 psi is the *maximum* pilot pressure needed for all valve sizes.

The direct operated directional valve *without* feedback, was designed to handle only small-to-moderate flows so the flow forces generated would not become excessively large. This valve does not, however, have the accuracy of the direct operated directional valves *with* feedback.

DIRECT OPERATED PROPORTIONAL RELIEF VALVES

Direct operated, pressure relief valves, Figure 1.26 *(a)*, are poppet type valves which are adjustable electronically with an LVDT for positional feedback, Figure 1.26 *(b)*.

The pressure setting of this valve is directly proportional to the input signal it receives. When the proportional solenoid is energized, the solenoid pin pushes directly on a pressure pad which acts on a

Fig. 1.26 (a) Direct operated proportional relief valve.

compression spring. The spring, in turn, exerts a force on the poppet. The actual position of the pressure pad is determined by the LVDT which supplies a feedback signal to the amplifier.

The feedback and input signals are compared electronically supplying a corrected signal back to the solenoid. A definite position of the pressure pad is then maintained. A very accurate spring tension is also maintained, accounting for very precise pressure settings. When operating pressure exceeds the setting of the spring force, the poppet opens, allowing oil to flow from the pressure port through the spring chamber, to tank. As was the case with the pilot operated proportional pressure relief valve, the pressure setting can be increased or decreased gradually with the help of the amplifier.

Since operating pressure can be set very accurately, this valve is used extensively in plastic injection molding machines where plastic

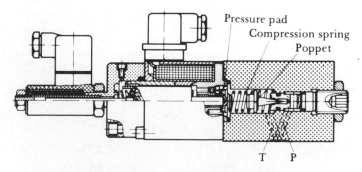

Fig. 1.26 (b) Cross-section through direct operated proportional pressure relief valve illustrated in Fig. 1.26 (a).

injection pressure accuracy is critical. This valve can also be used as a pilot control for logic elements and pressure relief valves. Maximum flow capabilities are limited by the various pressure ranges because seat diameter decreases as pressure capability increases.

If power is lost, solenoid force drops to zero and the valve pressure setting depends only on the unloading characteristics of the valve.

PROPORTIONAL PRESSURE COMPENSATED FLOW CONTROL VALVES

Proportional flow control valves are pressure compensated 2-port valves in which the main control orifice is adjusted electronically. Similar to conventional pressure compensated flow control valves, a proportional pressure compensated flow control valve maintains constant flow output by keeping the pressure drop across the main control orifice constant. The proportional valve, however, is different in that the control orifice has been modified to work in conjunction with a stroke controlled solenoid.

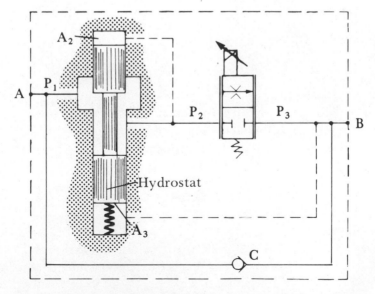

Fig. 1.27 Proportional pressure compensated flow control valve
(a) and operating circuit diagram (b).

In a 2-port proportional pressure compensated flow control valve, an electrically adjustable control orifice is connected in series with a pressure reducing valve spool, known as a hydrostat, Figure 1.27. The hydrostat is located *upstream* of the main control orifice and is held open by a light spring. When the input signal to the solenoid is zero, the light spring force holds the main control orifice closed. When the solenoid is energized, the solenoid pin acts directly on the control orifice, moving it downward against the spring, to open the valve and allow oil to flow from port A to port B.

At the same time, the LVDT (linear variable differential transformer) provides the necessary feedback to hold position. In this case, the LVDT provides feedback to maintain a very accurate orifice setting.

Pressure compensation is achieved by supplying a pilot passage from the front of the control orifice to one end of the hydrostat, A_2, and feeding a pilot passage beyond the control orifice to the opposite end of the hydrostat A_3, assisted by the force exerted by the spring. Load induced pressure at the outlet port or pressure deviations at the inlet port are thus compensated by the hydrostat, providing constant output flow.

The amplifier provides time controlled opening and closing of the orifice. For reverse free flow, a check valve C, built into the valve provides a flow path from port B to port A. Proportional flow control valves are also available with either linear or progressive flow characteristics. The input signal range is the same for both. However, the progressive flow characteristic gives finer control at the beginning of orifice adjustment.

In case electrical power or feedback is lost, solenoid force drops to zero and the force exerted by the spring closes the orifice. When feedback wiring is connected incorrectly or damaged, a LED (light emitting diode) indicates the malfunction on the amplifier card.

PROPORTIONAL FLOW LOGIC VALVES

Proportional flow control logic valves are basically electrically adjustable flow controls that fit into a standard logic valve cavity, Figure 1.28 *(a)*. The cover and cartridge are assembled as a single unit, with the cover consisting of a proportional force solenoid and a pilot controller, Figure 1.28 *(b)*.

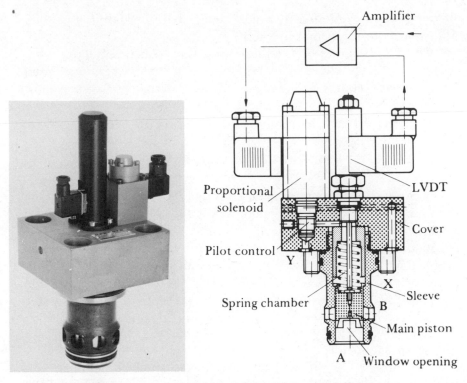

Fig. 1.28 (a) Proportional flow logic valve.   (b) Cross-section.

When an electrical signal is fed into an electronic amplifier, the solenoid and controller adjust the pilot pressure supplied from port *A* to change spool position. An LVDT then feeds back the position to the amplifier to maintain the desired orifice condition for flow from port *A* to port *B*. The proportional logic valve is available with either linear or progressive flow characteristics which are adjusted by a 0 to 6-volt, 0 to 9-volt, or differential ± 10-volt command signal.

Because the valve remains relatively unaffected by changes in system pressure, it can open and close the orifice in the same length of time. This maximum time can be changed on the amplifier card by adjusting a built-in ramp generator.

The amplifier can be used in several ways. An external potentiometer can make the orifice remotely adjustable while maximum spool acceleration is still limited by this internal ramp; or a limit switch can be added to turn the ramp on and off. In case of power failure, the element will return to its normally closed position.

PROPORTIONAL VANE PUMP

A proportional vane pump is essentially a variable displacement vane pump with the control operation of a load sensing control. The pump, therefore, delivers *constant* output flow as long as there is a constant pressure drop at the orifice. Also, when one compares this proportional vane pump with a standard vane pump with load sensing, both the throttle orifice and pressure relief valve (which is used for the pressure compensation stage) can be set electrically for flow and pressure control rather than setting these adjustments manually.

Basic Pump Operation

The cam ring of the variable volume vane pump is held between two control pistons. The larger piston, which is backed by a light force spring has twice the effective area of the smaller piston. Thus, during start-up, the spring force holds the cam in an *eccentric* position, as illustrated in Figure 1.29 *(b)*, allowing the pump to pump fluid.

System pressure is acting on both pistons. As long as the larger piston is *not* vented to tank, the pump will maintain output flow. As soon as the larger piston is vented to tank, the smaller piston, which is still pressurized, pushes the cam to a concentric, no-flow position. It will so remain until *both* sides of the piston are again pressurized.*

Control Operation

Pump output flows through electronically adjustable orifice *A*,

---

*For a more detailed explanation of variable volume pumps, see **Using Industrial Hydraulics** by *Tom Frankenfield, published by the Penton/IPC Co., Cleveland, Ohio.*

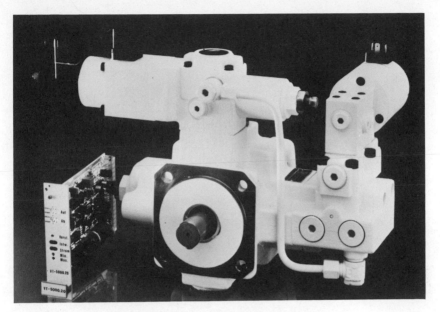

Fig. 1.29 (a) Proportional vane pump.

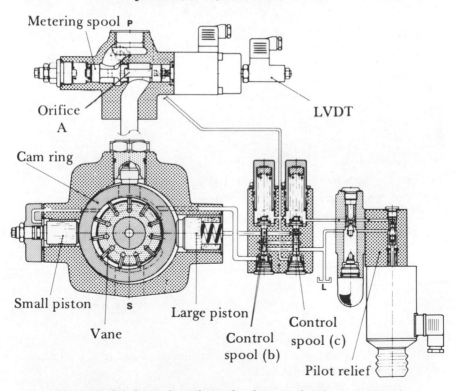

Fig. 1.29 (b) Operating schematic of proportional vane pump.

Figure 1.30. For any setting of orifice A, the pressure drop is held constant by the load sensing pump control. The control spool in the pump control works much like a hydrostat in a pressure compensated flow control valve. Control spool B, Figure 1.30, senses pressure at the inlet *and* outlet of the proportional orifice.

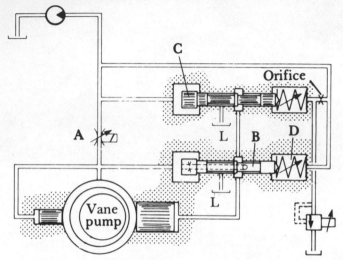

Fig. 1.30 *Control spool B in pump control works much like hydro-stat in pressure compensated flow control valve.*

The outlet sensing side of the spool is aided by an adjustable spring force which is normally set at about 150 psi. The setting of this spring force determines the pressure drop across the main orifice. By adjusting this differential pressure, accurate flow can be achieved for any given electronic input signal.

Since the control spool areas are equal at both ends, the spool modulates to maintain a balance between inlet pressure exposed on the left side and the combination of outlet pressure and spring force on the right. When the orifice shifts toward a closing direction, or when there is a loss in load induced pressure, there is a tendency for inlet pressure to exceed outlet pressure plus spring force. However, this cannot happen because the higher inlet pressure shifts the control spool to the right, partially unloading the pump's larger control piston. Pump flow decreases until inlet pressure again balances outlet pressure plus spring force.

Conversely, if the proportional orifice is open, or if load induced pressure increases, inlet pressure at the orifice is no longer sufficient to maintain pressure balance. However, this condition unbalances the spool so it shifts left, loading the pump control piston. Pump flow increases until the resistance to flow at the main orifice re-establishes the modulating pressure balance on the control spool. Because the electronically selected flow is influenced by only the orifice area, A, and the constantly maintained pressure drop, the pump's volumetric

$$Q \text{ (gpm)} = CA\sqrt{\Delta P}$$

efficiency does not affect the desired flow selection.

Proportionally Adjusted Pressure Compensation

The electrically adjustable pressure and pump control spool, C, Figure 1.30, works much like the electronic proportional relief valve discussed previously. As long as load induced pressure acting on the effective area of the pilot poppet does not exceed the adjusted solenoid force, there is no pilot flow, and therefore there is no differential pressure acting on the orifice located next to the spring chamber D. The spool is pressure-balanced and the spring force keeps the spool shifted to the left in a *pump-loaded* position.

When load induced pressure exceeds the setting of the proportional relief valve, pilot flow across the fixed orifice creates a pressure imbalance on control spool C. The spool snap-shifts to the right providing pressure compensation for the pump.

PROPORTIONAL PRESSURE RELIEF VALVES

The proportional pressure relief valve is direct-operated and controlled by a force solenoid. The desired amount of signal to the solenoid determines the maximum pressure at which the pump compensates. Likewise, the minimum pressure at which the pump compensates is determined by the setting of the spring adjustment on the pressure control spool, which is generally set at a low pressure for proper operation of the pump control.

The pilot valve can handle flows to 3.2 gpm, more than enough to drain oil from the control section of the pump. Sandwiched directly under the electronic proportional valve, is a mechanically adjustable pressure relief valve. This valve, which is similar to the proportional pilot operated relief valve with maximum pressure protection, can be adjusted slightly above the setting of the proportional force solenoid. In case of electrical power failure, or high current peaks, the pump always has maximum protection. The sensitivity of the solenoid requires that the tank port be connected separately back to tank to avoid any backpressure that might otherwise develop in the valve.

Electronically Adjustable Main Orifice

The main orifice is a spool type, variable orifice controlled by a proportional solenoid and an LVDT for positional feedback. In Figure 1.31, the orifice is shown closed. As the desired amount of

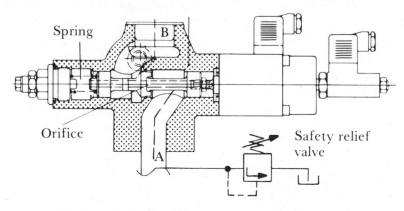

Spring   B

Orifice

A

Safety relief valve

Fig. 1.31 Schematic of electronically adjustable main orifice in proportional pressure relief valve.

signal is increased, the solenoid acts directly on the spool which is counteracted by a light spring at the opposite end of the spool. In case electrical power or feedback is lost, the force exerted by the spring will shift the spool back to a closed position. At each position of the spool, the orifice opens proportionally, allowing pump output to flow from port A to port B.

For maximum pump protection, a quick acting, mechanically adjustable relief valve can be installed in the orifice housing. This relief valve would be placed just before the inlet side of the orifice (A side) to avoid high pressure peaks.

REVIEW QUESTIONS

1.1   Are there any differences between servovalves and proportional valves?

1.2   If so, list 5 distinguishing characteristics of each.

1.3   On what factors does the performance of a proportional valve depend?

1.4   How many types of proportional solenoids are there? Name them and describe them briefly.

1.5   How does a conventional DC solenoid work?

1.6   How does a proportional force solenoid work?

1.7   Is there any relationship between current and force? Explain.

1.8   What is the maximum force output for force controlled solenoids?

1.9   What is a wet pin type solenoid? Discuss briefly.

1.10 Do proportional pressure relief valves perform the same functions as pilot operated types?

1.11 What is the major difference between the two types mentioned in question 1.10?

1.12 Explain how a conventional solenoid valve works.

1.13 Explain how a proportional directional control valve works.

1.14 Explain why the pilot head of a proportional relief valve must be drained externally.

1.15 What would happen if it were drained internally?

1.16 Does it provide any advantages? If so, which ones?

1.17 How long does it generally take to adjust the response time of a solenoid?

1.18 What happens to the solenoid force in case of electrical power failure?

1.19 Discuss the differences between proportional, pilot operated pressure reducing valves and proportional, pilot operated pressure relief valves.

1.20 Explain how they work.

1.21 What is an electronic amplifier? What does it do?

1.22 What is a proportional directional valve?

1.23 How does a proportional directional valve differ from a conventional directional valve?

1.24 Discuss the operating features of one over the other.

1.25 In a proportional directional valve, what is the diametral clearance between bore and spool? How does that compare to the diameter of a human hair?

1.26 What is a closed center spool configuration? What is an open center spool configuration? Draw the symbols for each.

1.27   When referring to hydraulic valves, what do the letters A, B, P and T stand for?

1.28   What is a restricted center spool?

1.29   In hydraulics, what is meant by regeneration?

1.30   What is the purpose of notches in a valve spool?

1.31   What is meant by a 1:1 or a 2:1 area ratio spool?

1.32   What is meant by a 1:1 or a 2:1 area ratio cylinder?

1.33   Referring to Fig. 1.14, explain how the 3-port proportional pressure control valve works.

1.34   What is a pilot operated proportional directional valve?

1.35   How can this be piloted?

1.36   Are pilot pressure requirements critical? Explain.

1.37   Referring to Fig. 1.17 (a) and (b), explain how a pilot operated, proportional directional valve works.

1.38   Is there any relationship between the magnitude of an electronic signal and spool response? Explain.

1.39   Discuss the differences between a direct operated proportional directional valve and a hydraulic pilot operated proportional directional valve.

1.40   What is meant by metering?

1.41   Generally, how many types are there? What are they? How do they differ?

1.42   What factor controls how fast an actuator moves?

1.43   What factor controls the size load an actuator can handle?

1.44   What is meant by maximum cycle rate?

1.45   Can computers be used in conjunction with hydraulic systems? Explain.

1.46   What is the connection between rate and productivity? Is this important? Why?

1.47   What is meant by cps? What is it a measure of?

1.48   What is meant by frequency response?

1.49   Define frequency, amplitude response, decibel.

1.50   What is phase lag? In what units is phase lag measured?

1.51   What is a Hertz? What is its abbreviation?

1.52   Distinguish between these three: (1) electronically controlled, pilot operated directional valves; (2) hydraulically controlled, proportional directional control valves; and (3) electronically controlled, direct operated proportional directional valve.

1.53   What is meant by stroke controlled solenoid?

1.54   What does LVDT stand for?

1.55   What do LVDTs do? How are they used?

1.56 What is feedback? What is its purpose?

1.57 Can feedback be used in proportional directional valves? For what purpose?

1.58 Is feedback in a valve necessary? Discuss.

1.59 What is a direct operated proportional relief valve?

1.60 What is a poppet type valve? What other types are there?

1.61 What is a directional control valve?

1.62 What is a pressure control valve?

1.63 What is a flow control valve?

1.64 What is pressure compensation?

1.65 Does pressure compensation have the same effect in a pump and in a flow control valve? Explain.

1.66 What are proportional flow logic valves? Are they known by another name? Which one.

1.67 What is meant by ramp (in an amplifier)?

1.68 What is load sensing?

1.69 What is a proportional vane pump?

1.70 Does load sensing help a pump supply constant output? How?

1.71 Are springs important in load sensing controls? Explain.

1.72 What is volumetric efficiency?

1.73 What is mechanical efficiency?

1.74 What is overall efficiency?

Chapter 2

SYSTEM DESIGN CONSIDERATIONS

When an engineer designs a hydraulic system for control with 4-port proportional valves, he must consider several important elements. This is especially true if the system will cycle heavy masses rapidly. One of the elements the engineer must take into account is the *natural frequency* of the system. From the laws of physics we know that the formula for undampened, natural frequency is:

$$\omega_0 = \sqrt{C/M}$$

where:

ω_0 - natural frequency

C - spring constant

M - moving mass $= w/g$

The spring constant for a hydraulic system can be related directly to the oil volume trapped between the 4-port proportional valve and the actuator. The moving mass is the weight of the object, w, being moved divided by gravity, g, which is equal to 32.2 ft/s^2.

The natural frequency of a hydraulic system is expressed in Hertz (Hz), and is a function of the mass and the oil volume trapped between the valve and the actuator. With this data you can determine how fast a load can be accelerated and decelerated without causing instability and subsequent damage to the system. To demonstrate this point, let us first consider a simple system, in which a weight w is attached to a spring. The natural frequency of the system is a function of the spring constant and the mass, Figure 2.1.

This frequency can be calculated mathematically to tell us how fast this weight can be moved back and forth without having the weight of the object directly oppose the input to the spring. For example, the input to the spring-mass system could be someone's hand moving the spring-mass system up and down a certain distance.

*Fig. 2.1 Natural frequency is a function
of mass and spring constant.*

As long as the spring is moved more slowly than the natural frequency of the total spring-mass system, the weight will follow the movement of the spring. There will be very little difference between the movement of the spring and the weight.

The faster the input, or hand movement to the spring, the more the weight will lag. If the input to the spring is at the same frequency as that of the total spring-mass system, as one's hand (the spring-mass system) moves down, the weight moves up. Likewise, as the hand moves up, the weight moves down. The weight would then be in direct opposition to the movement of the spring. This would result in a system performing a function opposite of that required. This is called instability, or resonance.

To put this concept in perspective with respect to a hydraulic system, the natural frequency can be calculated (as previously mentioned) and the factor which determines instability is the acceleration and deceleration time. Trying to accelerate or decelerate high inertial loads too quickly is likely to cause a cylinder or fluid motor to become unstable.

It is possible to increase the spring constant of the system by placing the valve as close to the actuator as possible; thus, in effect, reducing the oil volume trapped between valve and actuator. This arrangement allows higher acceleration and deceleration rates because the spring is stiffer. If, however, too high an acceleration rate is chosen, the actuator will have an irregular movement, regardless of the higher natural frequency. If the natural frequency is too low, the system will oscillate.

Now, let us calculate the natural frequency of a differential cylinder, Figure 2.2, and determine the acceleration time.

The first element to establish is the spring constant, C. The designer must be cautious when the spring constant is at its *minimum* value because this also means that the natural frequency is also at its

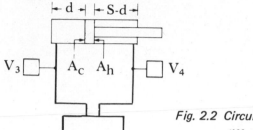

Fig. 2.2 Circuit diagram with basic differential area cylinder.

lowest value. This clearly is the *worst* spring constant condition and one *must* design around it.

Minimum spring constant, C_{min}, is a function of cylinder travel. An equation to calculate this distance has been developed for differential area cylinders. Here is the derivation for this equation.

To determine the distance at which the spring constant reaches its minimum value, the maximum value is first established by determining the spring constant with the cylinder fully retracted and fully extended.

When the cylinder is *fully retracted*, the bulk modulus equation can be applied, with stroke, $S = 0$:

$$\beta = (V_0 \cdot \Delta P)/\Delta V$$

$$\Delta P/\Delta V = \beta/V_0$$

where:

ΔP - change in pressure, psi

ΔV - change in volume, in^3

V_0 - original volume, in^3

β - bulk modulus, psi

The bulk modulus of a fluid is a measure of the change in volume which occurs when pressure on the fluid changes. The magnitude of volume change depends on the bulk modulus of the fluid, the original volume, and the amount of pressure change.

$$\Delta P = \Delta L / A_c$$

$$\Delta V = \Delta S \cdot A_c$$

where:

ΔL – change in load force, lb

ΔS – change in stroke, in.

A_c – area of cylinder cap end, in^2

The change in load is affected by acceleration force, $F = mA$.

Substituting,

$$(\Delta L / A_c) / (\Delta S \cdot A_c) = \beta / V_o$$

or

$$\Delta L / (\Delta S \cdot A_c^2) = \beta / V_o$$

To compute spring constant, C,

$$C = \Delta L / \Delta S = (A_c^2 \cdot \beta) / V_o$$

Thus,

$$C = (A_c^2 \cdot \beta) / V_o$$

This bulk modulus equation dictates the amount of ·pressure change for a given spring constant. Likewise the pressure change is created by the acceleration forces, $F = mA$

$$V_o = V_3 = (A_p L_1)$$

where:

V_3 - pipe volume in cylinder cap end, in^3

A_p - area of pipe or tube, in^2

L_1 - length of pipe or tube between valve and actuator cap end, in.

Thus,

$$C = (A_c^2 \cdot \beta) / V_3$$

The volume in the cylinder, the bulk modulus, and the volume between the valve and actuator head end, must all be considered because the proportional valve meters in *both* directions.

Thus:

$$C_{1\ max} = (A_c^2 \cdot \beta/V_3) + \left\{ A_h^2 \cdot \beta/[V_4 + (A_h \cdot S)] \right\}$$

V_4 - pipe volume at head end of cylinder, in^3

A_h - effective area of cylinder head end, in^2

S - stroke, in.

L_2 - length of pipe between valve and actuator head end, in

When the cylinder is fully extended,

$$C_{2\ max} = \left\{ (A_c^2 \cdot \beta)/[V_3 + (A_c \cdot S)] \right\} + [(A_h^2 \cdot \beta)/V_4]$$

The distance at which the spring constant is at a minimum is derived from:

$$dc/ds = 0 \quad i.e.,\ C' = 0 \text{ for } C' > 0$$

From the derivative, this distance equals:

$$d = [(A_h\ S/\sqrt{A_h^3}) + (V_4/\sqrt{A_h^3}) - (V_3/\sqrt{A_c^3})]/[(1/\sqrt{A_h} + (1/\sqrt{A_c})]$$

The curve for this cylinder would appear as in Figure 2.3

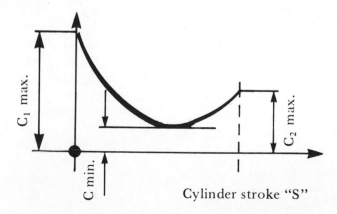

Fig. 2.3 Distance at which spring constant is at minimum.

$C_{min} = C_1 + C_2$ at stroke length, d

$C_{min} = \left\{ (A_c^2 \cdot \beta)/[V_3 + (A_c \cdot d)] \quad + \quad (A_h^2 \cdot \beta)/[V_4 + A_h(S-d)] \right\}$

$V_1 = V_3 + (A_c \cdot d)$

$V_2 = V_4 + [(S-d) A_h]$

where:

   V_1 - total oil volume on cylinder cap end side, in^3

   V_2 - total oil volume on cylinder head end side, in^3

Therefore, the natural frequency for the differential area cylinder would be:

$$\omega_O = \sqrt{C_{min}/M} = \sqrt{(C_1/M) + (C_2/M)}$$

$$\omega_O = \sqrt{[(A_c^2 \cdot \beta)/(V_1)] + [(A_h^2 \cdot \beta)/(V_2)]/M}$$

$$\omega_O = \sqrt{[(A_c^2 \cdot \beta)/(V_1 \cdot M)] + [(A_h^2 \cdot \beta)/(V_2 \cdot M)]}$$

As calculated above ω_O is the natural frequency in radians per second (rad./sec.). From a theoretical view ω_O can now be used to determine the time needed for acceleration. From experience, Figure 2.4, however, we must consider other capacitances of the system. (e.g. hoses, mechanical components, etc.). As proven by application examples, we find that the useable acceleration is better estimated by dividing the calculated natural frequency by 3. Naturally, this is a simplified estimate which has been proven to give ample accuracy for most systems. This simplification avoids a complex mathematical analysis which would require variables which are difficult, if not impossible to determine. Therefore; to calculate the useable acceleration:

$$\omega = \frac{\omega_O}{3} \text{ (rad./sec.)}$$

To obtain the natural frequency of the system in Hertz (Hz) we must divide by 2π.

$$F = \frac{\omega}{2\pi} \text{ (Hz)}$$

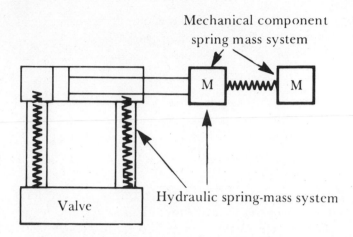

Fig. 2.4 System may be unstable if designed only around hydraulic spring-mass system.

Likewise; from ω (rad./sec.) we can also find the time constant t. This is the time period required for one oscillation.

$$t = \frac{1}{\omega} \text{ (sec.)}$$

As described in the following text, the time required for acceleration is based on this time constant. Generally, for stable acceleration, the time allowed must be a minimum of 4 to 6 times the time period for one oscillation. The mechanics involved are better described in Figure 2.5.

$t_b = 6 \times t$

t_b = Time allowed for acceleration (Figure 2.6)

t = Time period for one oscillation

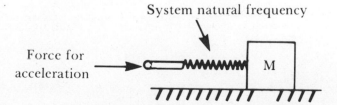

Fig. 2.5 Component forces of available input force.

During acceleration of any mechanical system, available force input is divided into three elements:

- a percentage of the force is used to cause the actual acceleration,

$$F = mA,$$

- a percentage of the force goes toward compressing the natural frequency of the system, $F = -Kx$, where K is the spring constant and x is the spring displacement (opposite of force), and

- some force is used to overcome frictional forces and other dampening factors.

For extremely low force levels, and thus low acceleration levels, the incoming force has a magnitude substantially lower than the compressive spring force of the system natural frequency. For these conditions, final velocity is predictable, being based entirely on the low acceleration rate and the desired change in velocity.

If, however, we attempt to increase input force to accelerate the mass, we reach a point where the spring (system natural frequency) *cannot* transmit this force. In other words, as input force increases, more of this force is used to compress the spring, while a limited, maximum force is transmitted through the spring to cause acceleration.

To obtain the maximum cycle rate for this system, we must supply enough force to accelerate the mass *without* supplying additional forces which only excite the natural frequency of the system, *i.e.*, compress the spring.

The ideal maximum acceleration is achieved when the extra force absorbed by the spring can be dampened quickly by the frictional forces (the dampening factor in the system). If spring forces are excessive when compared to dampening forces, an unstable oscillation is created.

Refering to the graphic analysis of maximum acceleration, Figure 2.6, the limiting acceleration is based on the time constant, as derived from:

$$V_f = V_d (1 - e^{-t/\tau})$$

where:

V_f - actual velocity

V_d - desired velocity

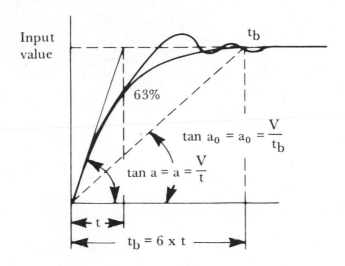

Fig. 2.6 Graphic analysis of maximum acceleration.

e - irrational number = 2.72

$-t/\tau$ - time constant based on number of time periods
     where t is the elapsed time and τ is a time constant

This relationship shows that during the *first* time period, maximum achievable velocity is 63% of desired velocity. During the second time period, velocity reaches 86% of desired; the third, fourth, fifth, and sixth time periods are 95, 97, 99, and 99+% respectively.

By allowing six time periods for acceleration, we achieve a smooth exponential increase to final velocity. If only four time periods are provided for acceleration, a critically dampened oscillation occurs *before* final velocity is reached. With less than four time periods, an unstable oscillation is created which, typically, cannot be tolerated. It is interesting to note that it does not matter what the acceleration rate is, the mechanical system always reaches 63% of desired velocity in the first time period. Likewise, it achieves a stable, constant velocity in the sixth time period.

As a practical matter, this information is important when selecting a system pressure and essential when establishing maximum pump flow. These considerations are illustrated in the following example. However, in an actual system, maximum acceleration is adjusted by starting with maximum ramp time

$$t_b = 6t$$

During machine operation, ramp time is gradually decreased until the instability point is reached. By adjusting ramp time to the point just before instability, (approximately six time constants) maximum acceleration, and therefore maximum production, can be achieved without shock.

At this point it is best to consider a conventional application, where a proportional directional control valve accelerates a load to a constant velocity and decelerates to a stop, then retracts in the same manner to start the cycle over again. Also, assuming the amplifier used has one ramp setting, acceleration and deceleration times will be the same.

Let us assume these parameters for the application: a cylinder with a 1½-in. bore with a 1-in. rod diameter must move a 1000-lb horizontal load a distance of 30 in. in 1 second.

$$w \ = \ 1000 \ lb$$

$$A_c \ = \ 1.76 \ in.^2$$

$$A_h \ = \ 0.98 \ in.^2$$

$$S \ = \ 30 \ in.$$

$$\beta \ = \ 2.0 \times 10^5 \ psi$$

Tube 3/4″ O.D. x .065″

L_1 - length of pipe at cap end = 46.5 in.
L_2 - length of pipe at head end = 38.75 in.

The natural frequency must be calculated first so that time to accelerate can be determined. From this value, velocity can then be determined.

Since the following information is known, and, as stated earlier, by designing around Cm, we can determine d, Figure 2.7.

$$d = \frac{[(A_h \cdot S)/\sqrt{A_h^3}] + [V_4/\sqrt{A_h^3}] - [V_3/\sqrt{A_c^3}]}{(1/\sqrt{A_h}) + (1/\sqrt{A_c})}$$

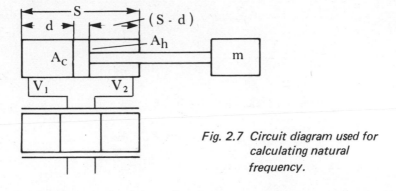

Fig. 2.7 *Circuit diagram used for calculating natural frequency.*

$V_3 = (\pi d^2/4)\, L_1 = (\pi \cdot 0.62^2/4)46.5 = 14.04$ in.^{3}

$V_4 = 0.30$ in.$^2 \cdot 38.75$ in. $= 11.7$ in.^{3}

$d = \Big\{ [(0.98\ \text{in.}^2 \times 30\ \text{in.}) / \sqrt{(0.98\ \text{in.}^2)^3}\,]$

$\qquad + 11.7\ \text{in.}^3 / \sqrt{(0.98\ \text{in.}^2)^3}\,] - [(14.04\ \text{in.}^3)$

$\qquad / \sqrt{(1.76\ \text{in.}^2)^3}\,] \Big\} / [(1/\sqrt{0.98\ \text{in.}^2}) + (1\ \sqrt{1.76\ \text{in.}^2})$

$\quad = 20.5$ in.

The natural frequency can now be calculated:

$\omega_o = \sqrt{[(A_c^2 \cdot \beta)/(V_1 \cdot M)] + [(A_h^2 \cdot \beta)/(V_2 \cdot M)]}$

$V_1 = V_3 + (A_c \cdot d)$

$\quad = 14.04$ in.$^3 + (1.76$ in.$^2 \cdot 20$ in.$) = 49.05$ in.^{3}

$V_2 = V_4 + [A_h(S - d)$

$\quad = 11.7$ in.$^3 + [0.98$ in.$^2\ (30$ in. $- 20$ in.$)] = 21.50$ in.^{3}

$$\omega_o = \sqrt{\dfrac{(1.76\ \text{in.}^2)^2 \times 2.0 \times 10^5\ \text{lbs/in.}^2 \times 32.2\ \text{ft/sec}^2 \times 12\ \text{in./ft}}{49.05\ \text{in.}^3 \times 1000\ \text{lbs}}} +$$

$$\sqrt{\frac{(0.98 \text{ in.}^2)^2 \times 2.0 \times 10^5 \text{ lbs/in.}^2 \times 32.2 \text{ ft/sec}^2 \times 12 \text{ in./ft}}{21.05 \text{ in.}^3 \times 1000 \text{ lbs}}}$$

$$\omega_o = \sqrt{4880/s^2 + 3526/s^2} = 91.4 \text{ radians/s}$$

The usable acceleration lies at about 1/3 of the natural frequency:

$$\omega = \omega_o/3 = 91.4/3 = 30.46/S$$

Acceleration time can now be calculated

$$t = 1/\omega = 1/30.46 = 0.032 \text{ sec.}$$

However, this value determines only the time in which the amplitude (which is equal to velocity/time) reaches about 63% of its desired final value. Acceleration time is proportional to final desired speed; therefore, a factor of $6t$ is used for acceleration stabilizing time. Using this factor of 6 for a proportional valve system has proven that this time acceleration and deceleration lies outside the unstable region.

Therefore, the acceleration stabilizing time would be:

$$t_b = 6t = 6 \cdot 0.032 = 0.20 \text{ sec.}$$

From this acceleration time, maximum velocity can be determined in terms of stroke. From V_{max}, acceleration, acceleration force, acceleration pressure, and required flow rate can be determined, Figure 2.8:

$$S = [2(V_{max}t_b)/2] + V_{max} \times (t - 2t_b)$$

$$V_{max} = s/(t - t_b)$$

$$= 30 \text{ in.}/(1.0 - 0.20) = (37.5 \text{ in./S})(60 \text{ S/min})$$

$$= 2250 \text{ in./min.}$$

$$a_{max} = V/t_b = (37.5 \text{ in./S}) / (0.20 \text{ S})$$

$$= 187 \text{ in./S}^2 = 15.6 \text{ ft/S}^2$$

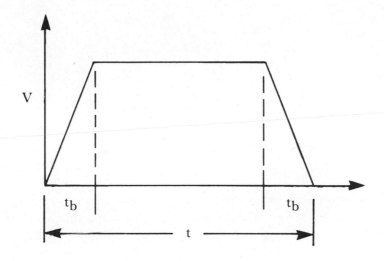

Fig. 2.8 Velocity vs. time relationship as a function of time periods.

Acceleration force would be:

$$F = ma = (w/g)\ a$$

$$= (1000\ lb/32.2\ ft/S^2)\ (15.6\ ft/S^2)$$

$$= 485\ lb$$

Frictional force:

$$F = \mu \cdot N = 0.58\ (1000) = 580\ lb.$$

$$F_t = 580\ lb + 485\ lb = 1065\ lbs.$$

Acceleration pressure at the cylinder cap end

$$P = F_t/A_b = 1065\ lb/1.767\ in.^2 = 605\ psi$$

Acceleration pressure at the cylinder head end would be:

$$P = F_t/A_h = 1065\ lb\ /\ 0.982\ in.^2 = 1085\ psi$$

Required flow rate would be:

$$Q_c = VA_c/231 = (2250\ in./min)\ (1.76\ in.^2)\ /\ (231\ in.^3/1\ gallon)$$

= 17.2 gpm

$$Q_h = VA_h/231 = (2250 \times 0.982)/231 = 10.0 \text{ gpm}$$

Since the cylinder has an area ratio close to 2:1, the valve selected should also have a spool area ratio of 2:1. From calculations to be shown later in the text, the pressure drop can be determined and the valve selected. If the valve were located closer to the cylinder, and the natural frequency recalculated, it would be seen that the natural frequency would increase. This would result in higher acceleration and deceleration rates, producing a faster cycle rate.

This relationship is shown in the following example. Note, also, that if pump flow was determined without considering acceleration time, the required one-second cycle time could not be achieved.

Natural Frequency for Motors

From the laws of physics related to motion, Figure 2.9:

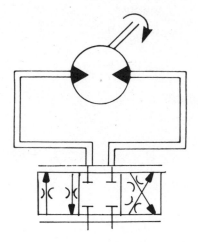

Fig. 2.9 Circuit used for calculating natural frequency of hydraulic motor.

$$\omega_0 = \sqrt{C/I}$$

where:

ω_0 - natural frequency

C - spring constant of the oil

I - moment of inertia of the mass

NOTE: Because the mass moment of inertia for rotary motion is a function of the object being rotated, an example for one application is shown. The effect of the proportional valve being located a substantial distance from the motor is also shown in this example.

Assume that in a molding machine, a carriage advances molding boxes to the molding line. A proportional directional valve controls a hydraulic motor which drives the carriage which is equipped with a pre-selected gear ratio, Figure 2.10. Here are the parameters of the system:

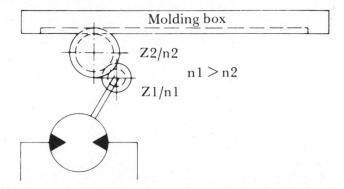

Fig. 2.10 *Fixed ratio gearing advances molding boxes in molding line.*

Each box weighs 8100 lb, must advance at a maximum velocity of 3.28 ft/sec, then stop within 1.5 ft.

Thus required acceleration, a, would be:

$$a = V^2/2S = (3.28 \text{ ft/sec})^2 \ / \ 2(1.5 \text{ ft}) = 3.56 \text{ ft/sec}^2$$

The gear ratio is:

$$Z_2/Z_1 = 38/17$$

Motor displacement = 6.7 in^3/rev

Desired motor speed = n_1 = 272 RPM

Tube diameter = 1/2 in. I.D.

Tube length = 32 ft

$$\omega_0 = \sqrt{C/I}$$

$$C_t = C_1 + C_2$$

$$C_{total} = [2(D/2\pi)^2 \cdot \beta]/V_1$$

where:

V_1 - trapped oil volume per side, in.^{3}

D - motor displacement, in.^{3}

β - 2.0 x 10^5 psi

I_t - mass moment of inertia

$V_1 = A_p (L + D/2)$

where:

A_p - pipe area, in.^{2}

L - pipe length, in.

$V_1 = [(\pi (0.5)^2/4)\ 393\ \text{in.}] + (6.7\ \text{in}^3/2)$

$\quad = 77\ \text{in.}^3 + 3.4\ \text{in.}^3 = 80.4\ \text{in.}^3$

Mass moment of inertia:

$I = (w/g)\ r^2$

where:

$w = 8100\ \text{lb}$

$r = 2.3\ \text{in. or } 0.2\ \text{ft}$

$I_t = I \dfrac{Z_1}{Z_2}$

$$I_t = (w/g)r^2 \times (Z_1/Z_2)$$

$$= [(8100 \text{ lb}/32.2 \text{ ft/sec}^2) (0.2 \text{ ft})^2 (17)^2] / (38)^2$$

$$= 2.1 \text{ lb-ft/S}^2$$

The natural frequency for rotary motion:

$$\omega_0 = \sqrt{2[(D/2\pi)^2 \cdot \beta] / (V_1 I_t)}$$

$$= \sqrt{2[(6.7 \text{ in.}^3/2\pi)^2 \cdot (2 \cdot 10^5 \text{ lb})]} \Big/$$

$$\sqrt{[80.4 \text{ in.}^3 (2.1 \text{ lb - f/S}^2) (12 \text{ in./ft})}$$

$$= 14.9 \text{ sec}^{-1}$$

$$\omega = \omega_0/3 = 14.9/3 = 4.96 \text{ sec}^{-1}$$

$$t = 1/\omega = 1/4.96 = 0.202 \text{ sec}$$

Acceleration time:

$$t_b = 6t = (0.202) 6 = 1.21 \text{ sec}$$

Acceleration rate:

$$a = V_{max}/t_b = (3.28 \text{ ft/S})/(1.21 \text{ S}) = 2.7 \text{ ft/S}^2$$

It was stated that the required acceleration was to be 3.56 ft/s^2. Calculations show that only a maximum acceleration of 2.7 ft/sec^2 is available for smooth running. To obtain a higher acceleration, the natural frequency of the system must be increased.

To increase the natural frequency of the system, the valve can be placed closer to the motor, reducing the volume of oil trapped between motor and valve. This arrangement provides a stiffer spring as well as higher acceleration and deceleration rates.

If the system is recalculated with a tube length of 3.5 ft, Trapped oil volume V_1:

$$V_1 = (A_p \cdot L) + (D/2)$$

$$= [(\pi (0.5)^2/4) (42 \text{ in.})] + (6.7 \text{ in}^3/2)$$

$$= 11.6 \text{ in.}^3$$

Natural frequency ω_0 is calculated from:

$$\omega_0 = \sqrt{2 \, [(D/2\pi)^2 \cdot \beta] / [(V_1 \cdot l_t)]}$$

$$= 40.9 \text{ sec}^{-1}$$

Achievable acceleration is:

$$\omega = \omega_0/3 = 40.9/3 = 13.65 \text{ sec}^{-1}$$

$$t = 1/\omega = 1/13.65 \text{ sec}^{-1} = 0.073 \text{ sec}$$

Acceleration time:

$$t_b = 6t = (0.073) \, 6 = 0.439 \text{ sec}$$

Acceleration rate:

$$a = (V_{max})/t_b = (3.28 \text{ ft/sec})/0.439 \text{ sec}$$

$$= 7.47 \text{ ft/sec}^2$$

With the reduced volume of oil between motor and valve, a recalculated acceleration rate of 7.47 ft/sec^2 now allows the load to be accelerated at 3.56 ft/sec^2 as the system remains stable.

Calculations of natural frequency in hydraulic systems with double rod end cylinders, Figure 2.11:

β -   bulk modulus of the oil, 2.0×10^5 psi

A_h -   head end cylinder area, in.^{2}

S -   cylinder stroke, in.

V -   total trapped oil volume, in.^{3}

m -   mass = w/g, lb/32.2 ft/sec^2

V_3 -   trapped oil volume in pipeline per cylinder side, in.^{3}

$$\omega_o = \sqrt{[(2 \cdot \beta \cdot A_h{}^2)/V(w/g)]} \ \text{sec}^{-1}$$

$$V = V_1 = V_2 = (A_h \cdot S/2) + V_3 \ \text{in.}^3$$

A double rod end cylinder is at its minimum frequency when it is a mid-stroke, S/2.

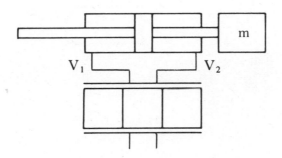

Fig. 2.11 Hydraulic system with double rod end cylinder.

Acceleration and Deceleration Curves

The families of curves illustrate acceleration and deceleration for *time*, Figure 2.12, and for *distance*, Figure 2.13 for constant acceleration. These curves may be used to make close estimates of required acceleration rates or to serve as a final check to confirm that previous acceleration calculations are correct.

Referring to the example again, where acceleration *time* was determined from the natural frequency for a cylinder with a piston area ratio of 2:1, acceleration time was 0.20 seconds, acceleration rate was 15.6 ft/sec², and velocity was 3.2 ft/sec. The acceleration time curves in Figure 2.12, show that constant acceleration will be achieved in 0.20 seconds at a velocity of 3.2 ft/sec, verifying that the calculations performed earlier were correct.

If, however, we arbitrarily choose a velocity and assume that we can accelerate a load in some particular time span, the curves also show that uniform acceleration will not be achieved. For example, assume that the load is to accelerate at 9 ft/sec² in less than 0.25 sec to a velocity of 3.3 ft/sec. *Constant* acceleration will not be achieved. The curve also shows that if a low acceleration rate is desired, a long acceleration time span will be needed. This can be

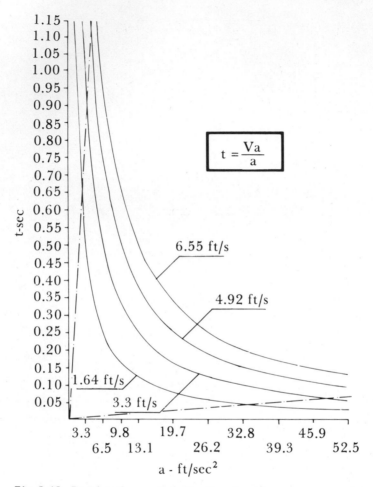

Fig. 2.12 Deceleration and acceleration time for constant acceleration.

established by the ramp setting of the amplifier. Ramp settings range from 0.03 to 5 seconds, more than enough to set acceleration and deceleration rates.

To determine acceleration *distance*, use the constant acceleration distance curves. Referring again to the same example, acceleration

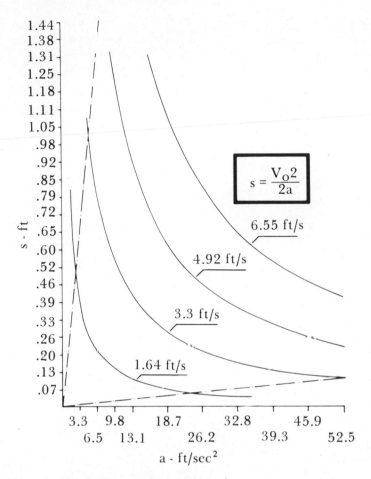

Fig. 2.13 Deceleration and acceleration distance for constant acceleration (deceleration).

distance is:

$$s = V^2/2a$$
$$= (37.5 \text{ in/sec})^2/(2 \cdot 187 \text{ in/sec}^2)$$
$$= 4.0 \text{ in. or } 0.33 \text{ ft}$$

When these values are analyzed graphically, you obtain constant acceleration, Figure 2.13. This distance can also be used as a starting point to establish where limit or proximity switches should be set. Because these switches are the electrical devices which a cylinder trips when acceleration or deceleration points are reached, they should be placed as close to the actual deceleration point as possible.

REVIEW QUESTIONS

2.1   What important elements must an engineer consider when designing a hydraulic system?

2.2   What is natural frequency?

2.3   What is spring constant?

2.4   What is mass?

2.5   What is gravity?

2.6   Is there any relationship between items 2.2 through 2.5 above?

2.7   What is acceleration? Deceleration? In what units are these expressed?

2.8   What is a spring-mass system? Discuss.

2.9   Discuss instability, resonance.

2.10   Do hydraulic systems have a natural frequency? Discuss.

2.11   What causes instability in a hydraulic system?

2.12   How can the spring constant be varied in a hydraulic system?

2.13   How does trapped oil volume affect spring constant?

2.14   What happens if a system's natural frequency is too high or too low?

2.15   What is bulk modulus?

2.16   How does bulk modulus affect natural frequency?

2.17   In a hydraulic system, how can the spring constant be changed?

2.18   Discuss the relationship between Figures 2.1 and 2.2.

2.19   Discuss the diagram in Figure 2.4.

2.20   What three elements make up available force input?

2.21   When is ideal maximum acceleration achieved?

2.22   What is a time period? How does it relate to velocity?

2.23   How many time periods are needed to reach final velocity?

2.24   What happens if less than four time periods are provided for acceleration?

2.25   Why is this information important in the design of a hydraulic system?

2.26   What is ramp time? Explain how it is used.

2.27   In algebra, how are parentheses, brackets and braces used in equations?

2.28   Do all mechanical components have a natural frequency?

2.29   What is mass moment of inertia?

2.30   What is constant acceleration?

2.31   Is it important in designing a hydraulic system? Explain.

Chapter 3

VALVE ANALYSIS

Calculating the natural frequency of a system is only one of the steps an engineer must consider. As a matter of fact, natural frequency need usually be considered only if *rapid* cycling is of *utmost* importance. When *accuracy* is the prime consideration, such as slow and smooth load movement, *estimated* natural frequency calculations may suffice.

The following discussion relates to two main types of proportional valves: those with spool area ratios of 2:1 and 1:1 and their effects on overrunning loads and resistive loads. From this information an accurate analogy can be made to select the proper valve for the application. This is the final design step when considering a proportional valve, regardless whether the natural frequency was calculated or estimated.

The following step-by-step calculations show how the equations were derived. These equations are tabulated at the end of this chapter for easy reference.

Overrunning Loads

Systems requiring cylinders with 2:1 area ratios should also be equipped with valves that have 2:1 area ratio spools. It was mentioned that a 2:1 area ratio spool is machined to provide half the flow area on one land as compared to the other. To further clarify this point, let us view mathematically why a valve with a spool with a 2:1 area ratio should be used with a cylinder which has a 2:1 area ratio.

All proportional valve spools can meter fluid in and out. Because of this orifice function, the equation for flow through an orifice applies:

$$Q = CA \sqrt{\Delta P}$$

where:

Q - flow across the orifice, gpm
C - discharge coefficient
A - area of orifice, in^2
ΔP - pressure drop across the orifice, psi

At first glance, it may seem that extensive calculations would be needed to determine the pressure drop across orifices ΔP_1 and ΔP_2, Figure 3.1. However, considering the load conditions of the system,

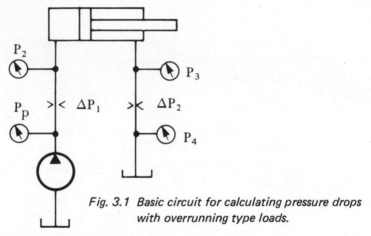

Fig. 3.1 Basic circuit for calculating pressure drops with overrunning type loads.

these calculations actually become quite simple. The first condition that can be satisfied is the orifice equation. This will first be done for a 2:1 area ratio spool, then for a 1:1 area ratio spool to show the adverse effects they cause when used with 2:1 area ratio cylinders.

For a 2:1 area ratio cylinder controlled by a valve with a 2:1 area ratio spool, Figure 3.2:

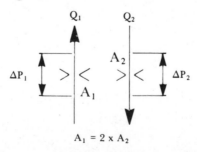

$$A_1 = 2 \times A_2$$

Fig. 3.2 Circuit for calculating pressure drops for system with 2:1 area ratio cylinders and 2:1 area ratio valves.

$$Q_1 = 2\,Q_2$$

$$A_1 = 2\,A_2 \text{ or } A_2 = A_1/2$$

Since there are two orifices:

$$Q_1 = CA_1\,\sqrt{\Delta P_1}$$

$$Q_2 = CA_2\,\sqrt{\Delta P_2}$$

Setting these equations equal to each other:

$$A_2 = A_1/2$$

Therefore,

$$A_2 = (Q_2/\sqrt{\Delta P_2}) = A_1/2 = (Q_1/\sqrt{\Delta P_1})\,/2$$

$$Q_1/2Q_2 = \sqrt{\Delta P_1}\,/\,\sqrt{\Delta P_2}$$

Thus,

$$\Delta P_1 \cong \Delta P_2$$

The pressure drop will be fairly equal on both sides of the valve, providing good controllability for 2:1 area ratio cylinders. However, for cylinders with 2:1 area ratios controlled by valves with 1:1 area

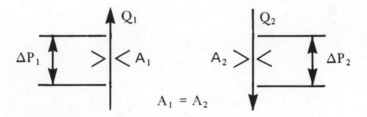

Fig. 3.3 Circuit for calculating pressure drops for system with 2:1 area ratio cylinders and 1:1 area ratio valves.

ratio spools, Figure 3.3, the situation will be different:

$$A_1 = Q_1/\sqrt{\Delta P_1} \text{ and } A_2 = Q_2/\sqrt{\Delta P_2}$$

Thus,

$$Q_1\,/\,\sqrt{\Delta P_1} = Q_2/\sqrt{\Delta P_2}$$

Therefore,

$$Q_1/Q_2 = \sqrt{\Delta P_1} / \sqrt{\Delta P_2}$$

$$4\,\Delta P_2 = \Delta P_1$$

When using a 2:1 area ratio cylinder with a 1:1 area ratio spool, ΔP_1 is four times greater than ΔP_2. This can cause substantial problems if the required backpressure on the head end of the cylinder must exceed 1/4 system pressure. A vacuum can be created because the cap end of the cylinder will not completely fill with oil. To study this condition in more detail, let us consider a situation with a 1000-lb overrunning load, a 1:1 area ratio spool, and a 2:1 area ratio cylinder, Figure 3.4. This relates directly to the method used to determine the pressure drop across the valve, Figures 3.1 and 3.2.

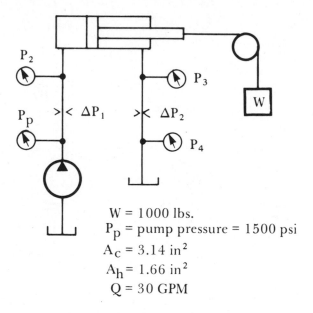

W = 1000 lbs.
P_p = pump pressure = 1500 psi
A_c = 3.14 in^2
A_h = 1.66 in^2
Q = 30 GPM

Fig. 3.4 Circuit diagram of potential overrunning load with 1:1 area ratio valve and 2:1 area ratio cylinders.

By summing forces, we solve for P_3.

$$-F = (P_2 \cdot A_c) - (P_3 \cdot A_h)$$

$$P_3 = [(P_2 \cdot A_c) + F]/A_h$$

We can then determine:

$$\Delta P_1 = P_p - P_2$$

$$\Delta P_2 = P_3 - P_4 \text{ ; assuming that } P_4 = 0,$$

$$\Delta P_2 = P_3$$

Determined previously for 1:1 area ratio spools. Squaring both sides:

$$Q_1/Q_2 = (\sqrt{\Delta P_1} / \sqrt{\Delta P_2}) \text{ or } (Q_1/Q_2)^2 = (\sqrt{\Delta P_1})^2/\sqrt{(\Delta P_2)}^2$$

$$= (Q_1)^2/(Q_2)^2 = \Delta P_1/\Delta P_2$$

$$\therefore \Delta P_2 = [(Q_2)^2 \cdot \Delta P_1] / (Q_1)^2$$

Substituting:

$$P_3 = [(Q_2)^2 \cdot (P_p - P_2)/Q_1{}^2]$$

$$(P_p - P_2)(Q_2)^2/(Q_1)^2 = (P_2 A_c + F)/A_h$$

Solving for P_2:

$$P_2 = [P_p(Q_2/Q_1)^2 - (F/A_h)] / [(A_c/A_h) + (Q_2/Q_1)^2]$$

$$= 1500 \text{ psi } (15 \text{ gpm}/30 \text{ gpm})^2 - 1000 \text{ lb}/1.66 \text{ in.}^2] /$$

$$[(3.14 \text{ in.}^2/1.66 \text{ in.}^2) + (15 \text{ gpm}/30 \text{ gpm})^2] = -106 \text{ psi}$$

When P_2 is less than zero psi, a vacuum is created.

$$\therefore \Delta P_1 = P_p - P_2$$
$$= 1500 \text{ psi} - (-106 \text{ psi}) = 1606 \text{ psi ?}$$

Since 1500 psi is max. ΔP_1;

$$\therefore \Delta P_2 = \Delta P_1/4 = 1500/4 = 375 \text{ psi}$$

A pressure drop of 375 psi across area A_2 is not enough. In other words, a smaller orifice is needed to create enough backpressure on the head end of the cylinder to keep the load from overrunning and causing a vacuum.

Next, let us consider the same condition with a 2:1 area ratio spool. The following relationships still hold:

$$P_3 = [(P_2 \cdot A_c) + F]/A_h$$

$$\Delta P_1 = P_p - P_2$$

$$\Delta P_2 = P_3$$

From the orifice calculation completed previously for a 2:1 area ratio:

$$2\,Q_2/\sqrt{\Delta P_2} = Q_1/\sqrt{\Delta P_1}$$

Thus,

$$\Delta P_2/\Delta P_1 = (2Q_2)^2/Q_1{}^2$$

$$\Delta P_2 = (2Q_2)^2 \cdot \Delta P_1/Q_1{}^2$$

Substituting:

$$P_3 = (P_p - P_2) \cdot (2Q_2)^2 / Q_1{}^2$$

Substituting again:

$$[(P_p - P_2) \cdot (2Q_2)^2] / Q_1{}^2 = (P_2 A_c + F)/A_h$$

Thus:

$$P_2 = [P_p (2Q_2/Q_1)^2 - (F/A_h)] / [(A_c/A_h) + (2Q_2/Q_1)^2]$$

$$= 1500 \, [(2 \cdot 15)/30]^2 - 1000/1.66$$

$$/[3.14/1.66] + [(2 \cdot 15)/30]^2$$

$$= 309 \text{ psi}$$

This cylinder will not draw a vacuum. Therefore,

$$\Delta P_1 = P_p - P_2 = 1500 - 309 = 1191 \text{ psi}$$

Since $\Delta P_2 = P_3$, and P_3 is known,

$$P_3 = \Delta P_2 = (P_2 A_c + F)/A_h$$

$$\Delta P_2 = (309) \cdot (3.14) + 1000 / 1.66 = 1182 \text{ psi}$$

The valve could now keep the load from overrunning and the system from pulling a vacuum. However, with a total pressure drop across the valve of 2373 psi, spool stroke would still have to be limited considerably. (Refer to the performance curves for the valve.)

At 70% control current, the total pressure drop across the valve is 1450 psi, Figure 3.5; the calculated pressure drop was 2373 psi.

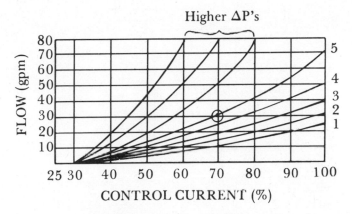

22.5 GPM nominal flow at 150 psi across valve

1 = 150 psi drop	4 = 725 psi drop
2 = 300 psi drop	5 = 1450 psi drop
3 = 450 psi drop	

Fig. 3.5 Family of curves showing pressure drop as a function of flow and control current signals.

This means that to obtain the 30-gpm flow required at a total pressure drop of 2373 psi, control current would have to be limited more than 70%. Because a small orifice is required at this considerably high pressure differential, very little of the spool stroke would be used. In addition, the performance of the valve at this high pressure differential will be poorer than in the range of 1450 psi and below at the required flow rate. The load should be counterbalanced. This subject is discussed later in the book.

Resistive Loads

Now that the conditions for 1:1 and 2:1 area ratio valve spools have been satisfied for overrunning loads, let us consider how 1:1 and 2:1 area ratio valve affect a resistive load, Figure, 3.6.

By summing forces, we can solve for P_2:

$$\Sigma F = (P_2 \cdot A_c) - [(P_3 \cdot A_h) + F]$$

$$P_2 = [(P_3 \cdot A_h) + F] / A_c$$

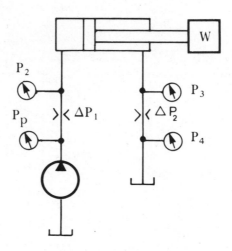

Fig. 3.6 Basic circuit for calculating pressure drops with resistive type loads.

For a 1:1 area ratio valve:

$$\Delta P_1 = (Q_1/Q_2)^2 \cdot \Delta P_2$$

$$\Delta P_1 = P_p - P_2 ; \Delta P_2 = P_3 - P\overset{0}{\diagup}_4$$

Substituting:

$$P_p - P_2 = (Q_1/Q_2)^2 \, P_3$$

$$P_p - (Q_1/Q_2)^2 \cdot P_3 = [F + (P_3 \cdot A_h)] / A_c$$

Thus,

$$P_3 = [P_p - (F/A_c)] / (Q_1/Q_2)^2 + (A_h/A_c)]$$

For 2:1 area ratio valve resistive load:

$$P_3 = [P_p - (F/A_c)] \,/\, [(Q_1/2Q_2)^2 + (A_h/A_c)]$$

Using the same parameters as before for a 2:1 area ratio valve with a resistive load:

$$P_3 = [1500 - (1000/3.14)] \,/\, [30/(2 \cdot 15)]^2 + (1.66/3.14)]$$

$$= 773 \text{ psi}$$

But, since $\Delta P_2 = P_3 = 773$ psi, then

$$P_2 = [(P_3 A_h) + F] \,/\, A_c$$

$$= [(773 \cdot 1.66) + 1000] \,/\, 3.14 = 727 \text{ psi}$$

Therefore,

$$\Delta P_1 = P_p - P_2 = 1500 - 727 = 773 \text{ psi}$$

$$\Delta P_t = \Delta P_1 + \Delta P_2 = 773 + 773 = 1546 \text{ psi}$$

The total pressure drop across the valve is 1546 psi. To obtain this pressure drop and use the maximum possible spool stroke, the required spool would have to be selected from the operating curves. However, a 2:1 area ratio spool is available only at its highest nominal flow rating for each valve size; therefore, in some cases, it may not be possible to use full spool stroke. This is especially true in the case of 2:1 area ratio cylinders and 2:1 area ratio spools. Comparing all operating curves for each valve size for a 2:1 area ratio spool, Figure 3.7, the best choice is a 22.5-gpm, 2:1-area-ratio spool for a required flow rate of 30 gpm and a total pressure drop of 1546 psi across the valve.

At 30 gpm and a pressure drop of 1546 psi, about 60% of control current is needed to establish the required total pressure drop. This considers only one direction of travel. If we recalculate the pressure drops in the opposite direction and assume that the cylinder will retract as fast as possible, the designer must make certain that flow and pressure drop requirements do not exceed the rating of the selected valve because flow from the cap end of the cylinder will be double that into the head end.

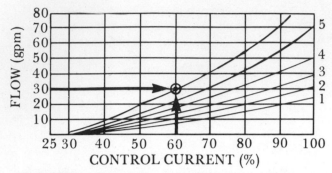

22.5 GPM nominal flow at 150 psi across valve

1 = 150 psi drop	4 = 725 psi drop
2 = 300 psi drop	5 = 1450 psi drop
3 = 450 psi drop	

Fig. 3.7 Family of curves showing pressure drop as a function of flow and control current signals.

In this example, the actual pressure drop would become quite high at a return flow of 60 gpm. However, not all applications require that the cylinder retract quickly. This means that the valve would be more than adequate if retract speed were not a major factor.

However, if the cylinder must retract as fast as possible, a larger 2:1 area ratio valve may have to be specified. Little spool stroke would be used in the extend mode, however, the amplifier presets can easily be set to provide the required speed.

For a cylinder with an area ratio that is close to 1:1, the equation for 1:1 area ratio valve is used. The parameters remain the same, except for the cylinder area ratio:

$A_c = 3.14$ in^2 and $A_h = 2.35$ in^2

$P_3 = 1500 - (1000/3.14) / [(30/22)^2 + (2.35/3.14)] = 454$ psi

$P_2 = [(454 \cdot 2.35) + 1000] / 3.14 = 658$ psi

$\Delta P_1 = P_p - P_2 = 1500 - 658 = 842$ psi

$\Delta P_t = 842 + 454 = 1296$ psi

Referring to Figure 3.8, consider an operating curve for a 13.21-gpm spool with a nominal flow rating of 150 psi across the valve: 30 gpm and a calculated pressure drop of 1296 psi; approximately 90% of the spool stroke can be used. For 30 gpm, the calculated pressure drop of 1296 will fall between curves 4 and 5.

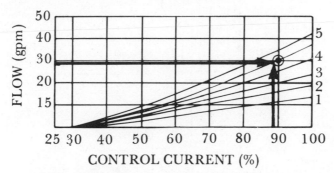

13.21 GPM nominal flow at 150 psi across valve

1 = 150 psi drop	4 = 725 psi drop
2 = 300 psi drop	5 = 1450 psi drop
3 = 450 psi drop	

Fig. 3.8   Operating curve for 13.21 gpm nominal flow at 150 psi pressure drop across the valve.

Counterbalancing with Proportional Valves

Proportional valves which control a vertical load which is raised and lowered by a cylinder with an area ratio close to 1:1, can use a direct operated counterbalance valve.

Although the proportional valve provides metering, the pressure drop needed at the required flow rate to keep the load from overrunning may become quite high. For example, consider a cylinder with an 8.29 in^2 cap end and a 6.80 in^2 head end which must move a 5000-lb load vertically, Figure 3.9. Flow requirements are 40 gpm and operating pressure is set for 1200 psi. Using the equation for overrunning loads, with a valve with an area ratio of 1:1:

$$P_2 = [P_p (Q_2/Q_1)^2 - (F/A_h)] / [(A_c/A_h) + (Q_2/Q_1)^2]$$

$$= [1200 (33/40)^2 - (5000/6.8)] / [(8.29/6.80) + (33/40)^2]$$

$$= 43 \text{ psi}$$

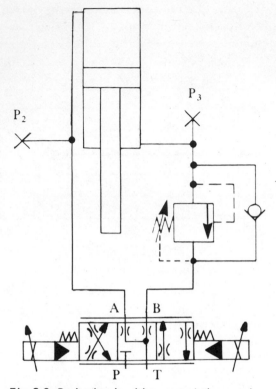

Fig. 3.9 Basic circuit with counterbalance valve.

$$\Delta P_1 = P_p - P_2 = 1200 - 43 = 1157 \text{ psi}$$

$$P_3 = [F + (P_2 \cdot A_c)] / A_h$$

$$= [5000 + (43 \cdot 8.29)] / 6.80 = 787 \text{ psi}$$

$$\Delta P_t = 1157 + 787 = 1944 \text{ psi}$$

A loop pressure drop of 1944 psi was calculated with no counter-balance valve. If we look at the particular pressure curve used, Figure 3.10, the loop drop is fairly high for the valve.

You can obtain better resolution with a direct-operated counter-balance valve. It may appear that the counterbalance valve should be used directly between the head end of the cylinder and the proportional valve. However, the designer must be aware of disadvantages to doing this.

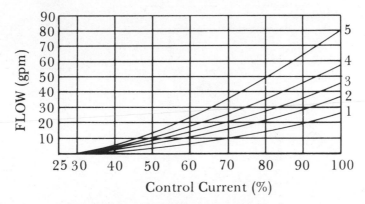

26.4 GPM nominal flow at 145 psi across valve

1 =	145 psi drop	4 =	725 psi drop
2 =	290 psi drop	5 =	1450 psi drop
3 =	435 psi drop		

Fig. 3.10 Family of curves showing pressure drops in circuit without counter-balance valve.

The setting of the counterbalance valve would be the pressure force acting on the cylinder piston area needed to keep the load suspended. Also, we must remember that the proportional valve adds resistance downstream of the counterbalance valve. The valve spring chamber is, therefore, increased to whatever the pressure drop is over the proportional valve from port B to port T.

In actuality, the proportional valve controls the cylinder as if there were no load because the relief valve holds the load at its pressure setting. Consequently, pressure on the load side of the cylinder can become quite high.

To demonstrate this point, let us consider a situation with the following parameters. A double rod end cylinder has an area of 10 in^2 and an overhung load of 45,000 lb, Figure 3.11. The actual setting of the counterbalance valve should be slightly more than 4500 psi.

Because the cylinder and valve are of equal areas, the pressure drop across the valve from port P to port A and from port B to port T will be shared equally. Also, since the counterbalance valve setting influences this condition, as if there was no load, by summing forces, the pressure drop on both sides of the valve will be 2500 psi.

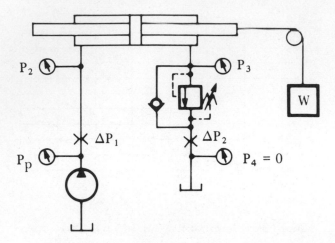

Fig. 3.11 Proportional valves controls cylinder as if there were no load because relief valve holds load at its pressure setting.

Forces will balance so that the pressure on the load side of the cylinder will be 2500 psi and 4500 psi, for a total of 7000 psi.

Likewise, pressure at the opposite end will be 2500 psi; 7000 psi at the head end is obviously too high and cannot be tolerated. In an actual application, if counterbalancing were used this way, depending on load conditions, pressure at the head end would not

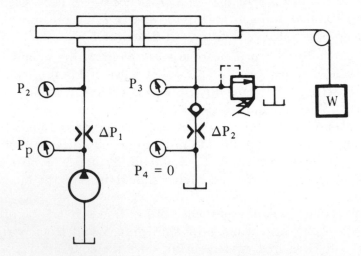

Fig. 3.12 By connecting counterbalance valve tank port directly to tank, proportional valve no longer affects relief valve setting.

be as high. The problem still exists, however, and the designer must be aware of what might happen. A more acceptable way of counterbalancing is shown in Figure 3.12.

By using a check valve and connecting the counterbalance valve tank port directly to tank, the proportional valve no longer influences the setting of the relief valve. Speed control can still be set electronically because the proportional valve will still be metering-in when lowering the load. This arrangement also prevents damage to the cylinder should the proportional valve close quickly because of power loss, because the counterbalance valve also functions as a relief valve.

When an application calls for a load to be held with a pilot-operated counterbalance valve, Figure 3.13, it is preferable to connect the valve tank port directly to tank. An example is locating the valve directly between the valve and the actuator where there are two restrictions in series.

Again, as with the direct-operated counterbalance valve, the pressure drop at the load side of the cylinder can become higher than expected because of the pressure felt at the spring chamber of the counterbalance valve. The designer must also remember that the setting of the counterbalance valve must exceed the pressure on the

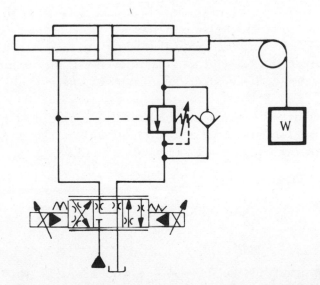

Fig. 3.13   When counterbalance and proportional valves are connected in series, internal drain port increases pressure setting of counterbalance valve.

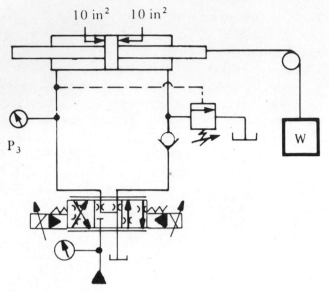

Fig. 3.14   When load is held with pilot-operated counterbalance valve,
connect valve tank port directly to tank.

no-load side of cylinder P_3.

In this case, as the load is being raised, it acts as a resistive load. Since the proportional valve *adds* resistance to flow at the no-load side of the cylinder, pressure at the no-load side may exceed the setting of the counterbalance valve when the load is being raised. Because this could cause erratic load movement, the counterbalance valve *must* be set higher than the pressure at the no-load side of the cylinder.

To calculate the pressure at P_3 for this condition, Figure 3.14, the equation for a 1:1 area ratio cylinder with a 1:1 area ratio valve can be used. See equations chart on page 87 and 88.

$$P_3 = [P_p - (F/A_r)] \; / \; [(Q_1/Q_2)^2 + (A_c/A_h)]$$

$$= [(5000 - 4500) \; / \; [(30/30)^2 + (10/10)]$$

$$= 250 \; psi$$

The pressure at the no-load side would be 250 psi; therefore, the counterbalance valve would have to be set slightly higher than 250 psi.

TABLE 3.1 – OVERRUNNING LOADS (2:1 VALVE)

APPLICATION	OVERRUNNING LOAD (2:1 VALVE)	
DIRECTION		
PRESSURE AT P_2	$$P_2 = \dfrac{P_p \dfrac{(2 \times Q_2)^2}{Q_1^2} - \dfrac{F}{A_h}}{\dfrac{A_c}{A_h} + \left(\dfrac{2 \times Q_2}{Q_1}\right)^2}$$	$$P_2 = \dfrac{P_p \dfrac{Q_2^2}{(2 \times Q_1)^2} - \dfrac{F}{A_c}}{\dfrac{A_h}{A_c} + \dfrac{Q_2^2}{(2 \times Q_1)^2}}$$
PRESSURE AT P_3	$$P_3 = \dfrac{F + P_2 A_c}{A_h}$$	$$P_3 = \dfrac{F + P_2 A_h}{A_c}$$
PRESSURE DROP ACROSS VALVE	$\Delta P_1 = P_p - P_2$ $\Delta P_2 = P_3$ $\Delta P_t = \Delta P_1 + \Delta P_2$	$\Delta P_1 = P_p - P_2$ $\Delta P_2 = P_3$ $\Delta P_t = \Delta P_1 + \Delta P_2$

TABLE 3.2 – OVERRUNNING LOADS (1:1 VALVE)

PRESSURE AT P_2	$$P_2 = \dfrac{P_p \dfrac{Q_2^2}{Q_1^2} - \dfrac{F}{A_h}}{\dfrac{A_c}{A_h} + \dfrac{Q_2^2}{Q_1^2}}$$	$$P_2 = \dfrac{P_p \dfrac{Q_2^2}{Q_1^2} - \dfrac{F}{A_c}}{\dfrac{A_h}{A_c} + \dfrac{Q_2^2}{Q_1^2}}$$
PRESSURE AT P_3	$$P_3 = \dfrac{F + P_2 A_c}{A_h}$$	$$P_3 = \dfrac{F + P_2 A_h}{A_c}$$
PRESSURE DROP ACROSS VALVE	$\Delta P_1 = P_p - P_2$ $\Delta P_2 = P_3$ $\Delta P_t = \Delta P_1 + \Delta P_2$	$\Delta P_1 = P_p - P_2$ $\Delta P_2 = P_3$ $\Delta P_t = \Delta P_1 + \Delta P_2$

TABLE 3.3 — RESISTIVE LOADS (2:1 VALVE)

APPLICATION	RESISTIVE LOAD (2:1 VALVE)	
DIRECTION		
PRESSURE AT P_3	$$P_3 = \dfrac{P_p - \dfrac{F}{A_c}}{\dfrac{Q_1^2}{(2 \times Q_2)^2} + \dfrac{A_h}{A_c}}$$	$$P_3 = \dfrac{P_p - \dfrac{F}{A_h}}{\dfrac{(2 \times Q_1)^2}{Q_2^2} + \dfrac{A_c}{A_h}}$$
PRESSURE AT P_2	$$P_2 = \dfrac{F + P_3 A_h}{A_c}$$	$$P_2 = \dfrac{F + P_3 A_c}{A_h}$$
PRESSURE DROP ACROSS VALVE	$\Delta P_1 = P_p - P_2$ $\Delta P_2 = P_3$ $\Delta P_t = \Delta P_1 + \Delta P_2$	$\Delta P_1 = P_p - P_2$ $\Delta P_2 = P_3$ $\Delta P_t = \Delta P_1 + \Delta P_2$

TABLE 3.4 — RESISTIVE LOADS (1:1 VALVE)

PRESSURE AT P_3	$$P_3 = \dfrac{P_p - \dfrac{F}{A_c}}{\dfrac{Q_1^2}{Q_2^2} + \dfrac{A_h}{A_c}}$$	$$P_3 = \dfrac{P_p - \dfrac{F}{A_h}}{\dfrac{Q_1^2}{Q_2^2} + \dfrac{A_c}{A_h}}$$
PRESSURE AT P_2	$$P_2 = \dfrac{F + P_3 A_h}{A_c}$$	$$P_2 = \dfrac{F + P_3 A_c}{A_h}$$
PRESSURE DROP ACROSS VALVE	$\Delta P_1 = P_p - P_2$ $\Delta P_2 = P_3$ $\Delta P_t = \Delta P_1 + \Delta P_2$	$\Delta P_1 = P_p - P_2$ $\Delta P_2 = P_3$ $\Delta P_t = \Delta P_1 + \Delta P_2$

REVIEW QUESTIONS

3.1 When is the natural frequency of a system an important factor to be concerned about?

3.2 When is it "safe" to use the *estimated* (instead of calculated) natural frequency of a system?

3.3 What is an overrunning load? Discuss.

3.4 What is a resistive load? Discuss.

3.5 Explain and discuss the impact of 2:1 and 1:1 area ratios in control valve spools and in cylinders.

3.6 What is pressure drop? Is pressure drop important in a hydraulic system? Explain.

3.7 Is pressure drop the same as pressure differential? What is the mathematical abbreviation commonly used for pressure differential? Does it make logical sense?

3.8 What is the pressure drop relationship when a 2:1 area ratio cylinder is used with a 1:1 area ratio valve spool?

3.9 Is a vacuum in the cap or head end of a cylinder desirable? Discuss and explain.

3.10 What is backpressure? Can it be created? How is it used?

3.11 Does the expression "drawing a vacuum" mean the same as "causing cavitation"?

3.12 Are some types of loads more apt to cause cavitation than others? Which?

3.13 Can valves be specified to avoid loads from overrunning?

3.14 What is an orifice? What does an orifice do?

3.15 Could a hydraulic system operate with zero pressure drop? Explain.

3.16 In a hydraulic system, what does flow control? What does pressure control?

3.17 In a hydraulic system, what one factor controls the pressure at which the system will operate?

3.18 What is the purpose of a counterbalance valve? Draw the ISO graphic symbol for a counterbalance valve.

3.19 Can a proportional valve act as a counterbalance valve? Explain.

3.20 What is a direct operated valve?

3.21 What other types are there, if any?

3.22 Counterbalance valves include check valves. What is their purpose?

3.23 Can a counterbalance valve be plumbed in a circuit more than one way? Explain.

Chapter 4

APPLICATION EXAMPLES

Proportional valves are available with various spool configurations, each with its own intended application. It is important to remember that all spools are essentially *closed* center when the designer considers the actual flow rating of the valve. As mentioned in Chapter 1 center conditions are mainly restricted bleed passages.

Closed Center Spools

A *closed center* spool is normally used with hydraulic motors; equal area, double rod end cylinders; and single rod end cylinders with an area ratio close to 1:1. Because a cylinder is fully closed, it is essential to include in the design of the circuit necessary load protection, namely pressure protection and anti-cavitation circuitry.

Although vacuum conditions and inertia related peak pressures can be eliminated by *ramping* the spool to its closed center position, it is not good design practice to rely entirely on this valve characteristic. Remember also that either malfunctioning electronics or power failures can cause the spool to snap back to its center position, resulting in potentially costly and damaging peak pressures and/or vacuum conditions.

Here is a good application example of a properly applied, closed center spool.

In this circuit, Figure 4.1, one relief valve is used with four check valves to provide cross port relieving for both directions of motor rotation. The two check valves with springs set to crack open at 7 psi, insure that oil from the relief valve outlet port remains in the hydraulic motor circuit before it is allowed to return to tank by flowing over the 45-psi check valve.

This circuit works properly because oil flow to and from the motor is theoretically equal. Note, however, that this safety circuit provides better pressure protection than anti-cavitation. Motors with poor volumetric efficiency or those that stop by *coasting* because of a relief valve set to open at a low pressure, can leave too much oil out of the circuit, causing cavitation damage to the motor.

In the circuit in Figure 4.1, there is no provision to provide make-up fluid to the motor. On the other hand, if this motor circuit were but one part of an entire system, tank flow from other, simultaneously occuring functions could be used to provide the small percentage of make-up oil needed.

T-blocked Center Position

If the motor circuit in Figure 4.1 were a single function system, it would not be possible to supply make-up oil to compensate for

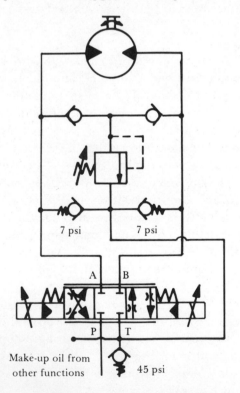

Fig. 4.1   Single relief valve is used with four check valves to provide cross-port relieving for both directions of motor rotation.

motor leakage. To prevent cavitation damage, consider using a spool, which, when in center position, pressurizes both working ports. Because this spool has a nearly *closed* center, it can provide only a small percentage of oil flow.

For double rod end cylinders and hydraulic motors, this small flow percentage is enough to compensate for leakage loss, Figure 4.2. However, it cannot supply enough oil to prevent cavitation condi-

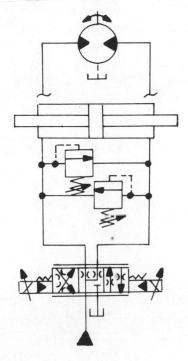

Fig. 4.2   For double rod end cylinders or hydraulic motors, a valve spool which pressurizes both working ports provides enough flow to compensate for normal leakage losses.

tions that would occur with the use of a port relief valve.

It is important to point out that only equal area actuators can function properly with cross port relief valve circuitry. A differential area cylinder produces unequal flows: when the cylinder extends, the cross port relief valve would relieve to the cap end of the cylinder, but could not supply enough oil to prevent a vacuum (cavitation) condition.

Likewise, when the cylinder retracts, the cross port relief would provide no pressure protection because the cylinder head end cannot accept all the oil relieved from the cylinder cap end. For this reason, the pressurized working port spool has little or no application potential with differential area cylinders. As in the previous example, Figure 4.2, cross port relief valves are generally recommended with equal area actuators.

A final note about this type spool is a word of caution in applications where an internally drained motor is specified. During normal operation, an internally drained motor drains leakage oil to its low pressure (outlet) port. This particular spool, however, pressurizes *both* ports simultaneously which could damage the motor and/or the shaft seal.

Bleed Center Spool

In its center (neutral) position, the bleed center spool connects the working ports to tank, Figure 4.3. The main purpose of this type spool is for use with differential area cylinders.

With this spool, internal leakage drains to tank, avoiding the possibility of pressure build-up when the proportional valve is

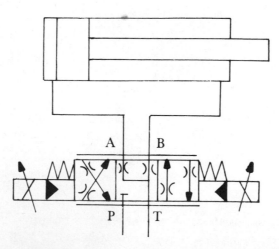

Fig. 4.3   In center position, a bleed center spool connects working ports to tank and is intended primarily for controlling differential area cylinders.

centered. Regardless whether conventional or proportional directional valves are used, it is generally recommended that designers NOT use fully closed center position spools with differential cylinders which have large diameter rods.

Remember that with closed center spools, there is always the possibility of having intensified pressure fluid build up in the cylinder, especially if the valve is left in center position for long periods of time. Such internal valve leakage can cause a cylinder to extend inadvertently because of the creation of a regenerative circuit. Likewise, intensified pressure on the cylinder head end can cause recurring rod seal failures.

Also note that the bleed center spool is *not* intended to function as a *float center* spool, normally associated with conventional directional control valves. The bleed center spool, for instance, cannot pass the entire flow of a counterbalance valve as it decelerates a load to a stop. Likewise, it cannot avoid cavitation by drawing oil from the tank through the orificed center design of the spool. To prevent vacuum conditions and inertia pressure spikes, additional circuitry is needed.

Figure 4.4 illustrates the typical circuitry required to avoid vacuum (cavitation) conditions and inertia pressure spikes when using differential area cylinders. With this type cylinder, port relief valves must be used instead of cross port relief valves.

Relief valve *A*, Figure 4.4, relieves full flow from the cylinder cap end to tank when the cylinder is under inertial load, as it retracts. Full make-up flow is provided to the cylinder head end over check valve *D* which is set to open at a very low pressure.

Likewise, check valve *C* and relief valve *B* provide a similar function when the cylinder extends. It is important to note that a small vacuum is all that is available to provide make-up flow to the cylinder. In circuit design, and for proper operation, the engineer must also consider the cracking pressures of the make-up check valves, line sizes, and, where applicable, oil head.

Regenerating 2:1 Ratio Cylinders

As was stated earlier, the bleed center spool is intended for use with differential area cylinders. If proportional speed control is needed for a regenerative circuit, several options are available.

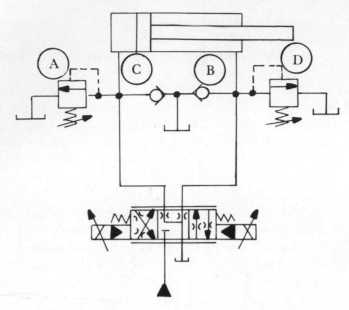

*Fig. 4.4   Example of circuit to avoid vacuum (cavitation) conditions and
inertia pressure spikes when using differential area cylinders.*

Figure 4.5 illustrates a typical circuit for cylinder regeneration. Unlike a conventional directional valve circuit, the proportional valve allows infinitely adjustable regenerative speed. This speed control is established by the orificed area from ports P to A. Notice that additional valving (not shown) would be required to take the cylinder out of regeneration to achieve maximum cylinder thrust.

Figure 4.6 shows a regenerative spool option which blocks ports B to T when the valve spool shifts after solenoid (b) is energized. This type spool simplifies plumbing and eliminates one check valve.

Regeneration with Counterbalancing

Often, it is necessary to regenerate a cylinder while simultaneously supporting an overrunning load. This can be achieved simply by replacing one of the check valves (see Figures 4.5 and 4.6) with a

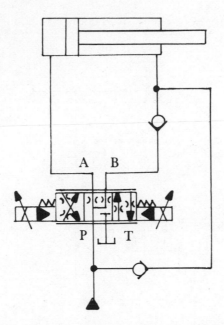

Fig. 4.5   Circuit illustrates cylinder regeneration.

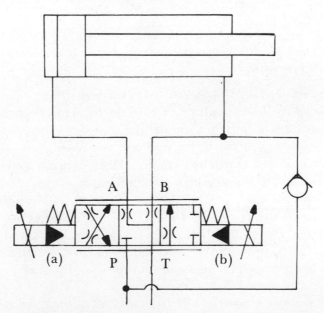

*Fig. 4.6   Regenerative spool option blocks port **B** to tank port when solenoid (**b**) is energized.*

relief valve, Figure 4.7. The desired counterbalance pressure can then be set on the valve.

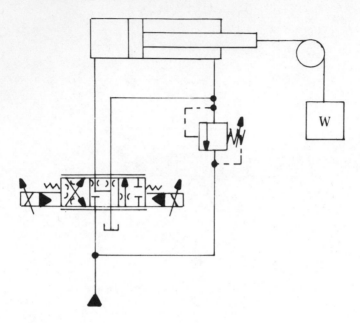

Fig. 4.7   By substituting relief valve for check valve in Fig. 4.6, cylinder is regenerated while system also supplies pressure fluid to overrunning load.

As mentioned in Chapter 1, the position of the main spool is proportional to the magnitude of the control signal voltage. Because spool position is infinitely variable, the designer is no longer limited to only a 3-position valve. By properly machining the spool, 4- or even 5-position valves can be obtained. The schematic drawing of the regenerative spool, Figure 4.8, is a typical example.

Proportional valve technology has substantially simplified regenerative circuitry. The circuit in Figure 4.8, shows a 4-position proportional valve and a direct operated relief valve. Circuit features include acceleration and deceleration control, infinitely variable speed, counterbalancing, and smooth transition between regenerative and full force cylinder extension.

If we consider how the circuit works when the cylinder extends, we see two separate operating modes. For low level commands from the amplifier card, position 3 valve C opens two orifices proportionally:

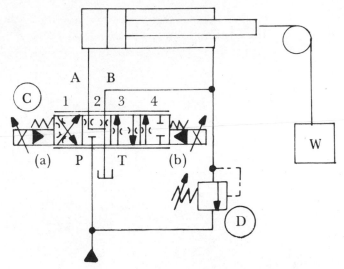

Fig. 4.8 Circuit illustrates use of 4-port, 4-position proportional directional valve and direct operated relief valve with overrunning load.

port P to port A and port B to tank. This arrangement provides proportional speed control in a full force mode. Because oil from the cylinder head end returns to tank through the proportional valve, maximum force is available for acceleration to desired speed. Likewise, returning to a low level command at the end of the stroke provides maximum force for the work portion of the cycle, for such tasks as pressing, clamping, etc.

Assuming that the cylinder is accelerating as it extends, once the command voltage is ramped to about 40% of its maximum value, the proportional valve shifts to its 4th position.

In this position, port B is blocked from port T, forcing oil to flow from the cylinder head end over the counterbalancing relief valve, D thus joining pump supply flow at port P of the proportional valve. The cylinder then moves smoothly into a regenerative operating mode. From 40% to 100% signal, cylinder speed is proportionally adjustable.

The purpose of a counterbalance valve is to provide braking when the system must decelerate from its relatively high regenerative speed. Basically a counterbalance valve prevents a load from running away.

As mentioned in the section on counterbalancing with proportional valves, when the command drops below 40%, counterbalancing and braking are provided only by the orificing characteristics of the proportional valve spool.

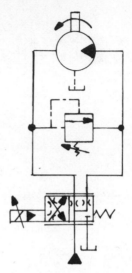

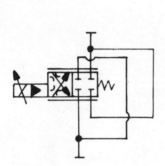

Fig. 4.9   Symbol shows how function of 2-position proportional valve differs from 2-position directional control valve.

Fig. 4.10   Circuit illustrates use of proportional valve to control speed of unidirectional hydraulic motor.

**Flow Control with
Proportional Valves**

As the fluid power symbols in Figure 4.9 show, a 2-position proportional valve differs in function from that of a 2-position directional control valve. Rather than provide *forward* and *reverse*, 2-position proportional valves are usually *double flowed* to provide infinitely variable speed control for only one direction of operation. Since the valve remains small, because of the double flow path, this type of electronic throttle valve is often an economical alternative to the more sophisticated pressure-compensated proportional flow controls.

In this application, however, we must be careful not to exceed the maximum tank port pressure rating of the valve. Generally speaking, direct-operated proportional valves can operate at pressures to 2000 psi; whereas externally-drained, pilot-operated valves can operate at pressures to 3600 psi.

When a designer chooses to use a 2-position proportional valve as electronically set throttles, it is not necessary that the valve always be double flow. The circuit in Figure 4.10 illustrates a proportional valve used to control the speed of a unidirectional motor. The orifices created by the valve meter oil both into and out of the hydraulic motor.

In the circuit in Figure 4.10, the unidirectional motor can be ramped to a desired (adjustable) maximum speed. Likewise, it can be decelerated at a separately adjustable ramp time. As discussed earlier in this Chapter, the spring offset position was selected to avoid motor cavitation as the braking force of the cross port relief valve decelerates the motor's inertial load to a stop.

REVIEW QUESTIONS

4.1   What is a proportional valve? Might one call it an analog or digital device? Explain.

4.2   Is a proportional valve a variable device? Explain.

4.3   What is a closed center spool? Draw an ISO graphic symbol to illustrate.

4.4   What is an open center spool? Draw an ISO graphic symbol to illustrate.

4.5   When should a designer consider using a closed center spool? An open center spool? Explain.

4.6   What is a uni-directional hydraulic motor? Bidirectional? What type is illustrated in Figure 4.1? Why was it used?

4.7   In Figure 4.1 explain the purpose of using four check valves with one pressure relief valve.

4.8   In Figure 4.1, why do two check valves have springs and two do not.

4.9   Does the same volume of oil leave the motor as enters the motor? Explain.

4.10   Can a hydraulic motor be made to cavitate? Does it matter? Explain.

4.11   What is make-up fluid? When is it used? Why?

4.12   What is crossport relief valving? When and why is it used?

4.13   The area of a cylinder cap and head ends differ in differential area cylinders. Could this be a factor in designing a system?

4.14   What does flow control in a hydraulic system?

4.15   What does pressure control in a hydraulic system?

4.16   What is meant by pressure intensification? What causes it?

4.17   What is internal leakage? Are all components susceptible to internal leakage?

4.18   In a hydraulic system, what is meant by fluid regeneration. Is this a desirable condition? Why? Explain.

4.19   What is a float center spool?

4.20   What is a bleed center spool?

4.21   What is the difference between crossport relief valving and port relief valving?

4.22   Can fluid regeneration be used with cylinders? With hydraulic motors? Discuss.

4.23   Is it possible to achieve fluid regeneration *and* counterbalancing of a load simultaneously? Explain.

4.24   Referring to Figure 4.8, explain how proportional valves have helped simplify regenerative circuitry.

4.25 What is a bleed center spool? What is one of its primary intended uses?

4.26 Referring to Figure 4.5 explain how a proportional valve provides infinitely adjustable regenerative speed control.

4.27 Does a regenerative circuit affect the thrust a cylinder can generate? Explain.

4.28 Because spool position is infinitely variable, the system designer is no longer limited to using only 3-position valves. True or false. Explain.

4.29 What happens to counterbalancing and braking when command drops below 40%?

4.30 What is meant by the statement: "proportional valves are usually *double flowed*." Is this a desirable characteristic? Explain.

Chapter 5

LOAD COMPENSATION

The electronic proportional valve is unique in its ability to adjust flow, or rate of change in flow, as quickly and frequently as necessary. However, for a given level of input, the valve alters only its orifice area. Flow through these orifices varies as the square root of the pressure drop. If a circuit requires precise speed control, regardless of speed variation, it becomes necessary to provide load compensation for the orifices.

The principle of load compensation is simple: it maintains a *constant* pressure drop across the orifice or orifices of the proportional valve. A pressure reducing or relieving function regulates the pressure at the orifice inlet to maintain this pressure at a constant differential when compared to outlet pressure.

Basic Requirements

Any load sensing (pressure compensated) function works on the principle that the sum of the forces generated by outlet pressure and spring opposes and modulates inlet pressure. Although this principle is well known in hydraulics, special consideration must be made when applied to proportional, 4-port valves.

In Chapter 4, we showed how to specify valves with respect to system pressure and load conditions. Figure 5.1 illustrates the relationship between flow, pressure drop, cylinder area, and piston velocity. The circuit must be designed so that all conditions are satisfied under all modes of operation and load conditions. There must also be a force balance at constant velocity.

If, for instance, a cylinder extends with a flow of 10 gpm across orifice A, at a compensated pressure drop of 100 psi, flow from the cylinder head end is reduced by the differential area (A_h/A_c), Figure 5.1. The lower flow means that there must be a smaller pressure

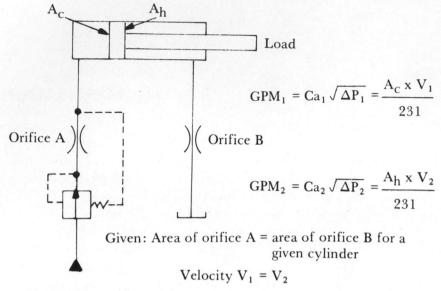

$$GPM_1 = Ca_1 \sqrt{\Delta P_1} = \frac{A_c \times V_1}{231}$$

$$GPM_2 = Ca_2 \sqrt{\Delta P_2} = \frac{A_h \times V_2}{231}$$

Given: Area of orifice A = area of orifice B for a
given cylinder

Velocity $V_1 = V_2$

*Fig. 5.1 Schematic illustrates system relationship between flow, pressure
drop, cylinder piston areas, and piston velocity.*

differential (1:1 area ratio valves) or an equal pressure differential
(2:1 area ratio valve *and* cylinder) at orifice *B*.

   If an overhung load or a decelerating force creates a pressure in
the cylinder head end in excess of 100 psi, the cylinder extends
faster than the amount of oil being supplied to the cylinder cap end,
creating a vacuum (cavitation) condition. This case points to the
need for a counterbalance or overcenter counterbalance valve between
the cylinder head end and the proportional valve. Refer to *Counterbalancing with Proportional Valves*, Chapter 3, page 81.

Selecting Downstream Pressure Signal

   The hardware used to load-compensate a proportional valve
includes conventional relief valves, reducing valves, sandwich-mounted
compensator plates, and logic cartridges. There is also a choice of
restrictive meter-in, restrictive meter-out or bypass flow regulation.
When designing circuits with pressure compensated proportional
valves, make certain that the compensating valve senses pressure at
the outlet of the orifice. This condition may preclude the use of
shuttle valves in a pressure feedback line.

METER-IN PRESSURE COMPENSATOR

A meter-in type pressure compensator consists of a sandwich type valve which can be mounted directly between a subplate and a proportional directional valve, Figure 5.2. It comes in two models. One has a feedback pressure port for custom designing; the other has a built-in shuttle valve between proportional valve ports A and B.

With this valve, pressure compensation is achieved without external pilot lines, assuming load conditions and circuits are proper and appropriate. Both are basically spool type, pressure-reducing valves. When sandwiched between the proportional directional valve and the subplate, they provide a *constant* output flow from port P to port A (or port P to port B) as required by the setting of the proportional valve.

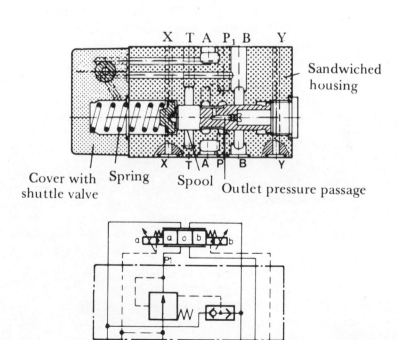

Fig. 5.2 Meter-in pressure compensator is sandwiched between subplate and proportional valve.

The compensator consists primarily of a housing, control spool, spring with its loading plate, and cover, Figure 5.2. A light spring holds the spool in open position. The spring exerts a force equivalent to about 120 psi acting on the area of the spool as P connects to $P1$.

This open condition is maintained as long as the differential pressure across the proportional valve (which acts as the main flow orifice) is *less* than 120 psi. As soon as the differential pressure attempts to exceed 120 psi, the control spool shifts to the left, reducing pressure at the inlet port of the proportional valve to modulate a 120-psi pressure drop across the valve, thus maintaining a constant output flow.

An increase in load-induced pressure felt at the outlet of the proportional valve shifts the spool to the right, increasing the pressure at the inlet of the proportional valve to re-establish the required 120-psi pressure drop across the valve. Likewise, a loss of load-induced pressure lowers the opening forces on the compensator spool. The spool then shifts in a closing direction, restricting inlet flow and maintaining the 120-psi pressure differential constant.

Note, too, that if inlet pressure increase rapidly, fluid pressure is felt on the right end of the spool, pushing it hard momentarily toward the left. This action opens port $P1$ to tank, draining oil until an equilibrium is re-established between inlet and outlet, thus avoiding high pressure peaks that might otherwise develop in port A.

Pressure Compensated Meter-in Sandwich

Generally speaking, meter-in pressure compensation is used in remotely adjusted or automated circuits where the designer wants to maintain a constant, but electronic, proportional speed. In this circuit, remember that the load must be accelerated to this speed. Likewise, the load must eventually decelerate to a stop. For any meter-in circuit, deceleration (and overhung load conditions) require the use of a counterbalance valve.

The circuit in Figure 5.3 shows a typical application of a meter-in, pressure compensator plate with built-in shuttle valve. Port relief valves protect the system from load-induced pressure spikes and provide counterbalancing as well as braking forces. Because these

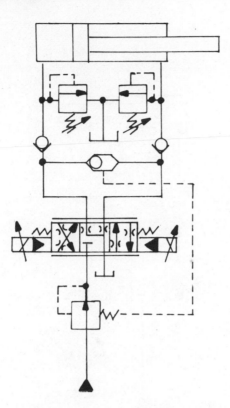

*Fig. 5.3 Typical application of meter-in pressure compensator plate with
builtin shuttle valve.*

relief valves bypass the flow orifices directly to tank in the proportional valve, the proportional valve controls speed only by controlling
meter-in flow.

In either shifted position of the proportional valve, the shuttle
valve selects the pressure feedback signal of the in-feeding oil. The
opposite side of the shuttle is vented to tank through the unused
passage of the proportional valve. Since this circuit is shown with
internally piloted counterbalance valves, overhung load forces should
be relatively constant.

Conventional Reducing Valve Used
as Meter-in Pressure Compensator

The circuit in Figure 5.4 is essentially the same as in Figure 5.3
except that it uses overcenter counterbalance valves for variable

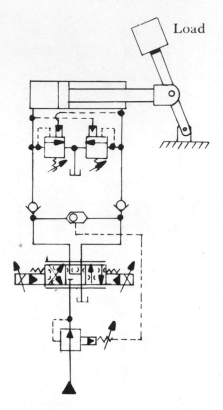

Fig. 5.4 Circuit is similar to that in Fig. 5.3, but uses two pilot operated, overcenter, counterbalance valves for variable load conditions and conventional pilot operated reducing valve for load compensation.

conditions and a conventional pilot operated reducing valve for load compensation.

When using a pilot operated reducing valve for load compensation, the valve is connected in series with port P of the proportional valve. Rather than connecting the return drain line from the reducing valve to tank, as you would in a conventional reducing valve circuit, the drain connection is shuttled to ports A and B of the proportional valve.

In operation, this circuit senses the in-feeding pressure in the pilot spring chamber of the pressure reducing valve. This combination of load induced pressure plus pilot spring force establishes the reduced pressure setting at the inlet of the proportional valve. A desirable characteristic of this circuit is that the pressure differential on the proportional valve can be adjusted by setting the pilot valve of the reducing valve. In this manner, for a given level of input to the proportional valve, an exact flow rate (or actuator speed) can be established.

Meter-in Pressure Compensation
without Counterbalancing

With highly resistive frictional loads, it is sometimes possible to meter-in fluid without counterbalancing. By electronically ramping deceleration, it is often possible to rely on the load frictional forces for deceleration. However, be careful when selecting a pilot signal for this type circuit.

Figure 5.5 illustrates a circuit where a meter-in compensator plate with a built-in shuttle valve can be used. Because the circuit has

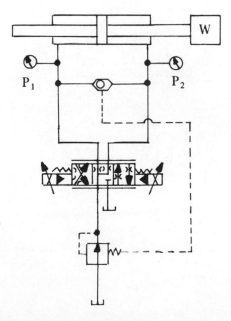

Fig. 5.5 In this circuit, meter-in compensator plate has built-in shuttle valve.

an equal areas cylinder, and braking forces are available due to a high level of friction, the shuttle valve compensator can be used successfully.

Assuming the cylinder moves to the right, load conditions dictate that *P1* will exceed that at *P2* during acceleration and constant velocity modes of operation. Also, because braking forces are purely frictional, pressure at *P2* may equal that at *P1* during deceleration, but it will never be higher.

Likewise, when the cylinder piston moves to the left, pressure at *P2* will always be the higher pressure. The circuit load conditions and actuator, as described above, allow the shuttle valve feedback of pilot pressure to operate properly. In other words, the shuttle valve is always providing feedback of the actual load-induced pressure at the outlet of the orifice. This fact, however, is seldom true when differential area cylinders are used. The circuit in Figure 5.6 illustrates this point.

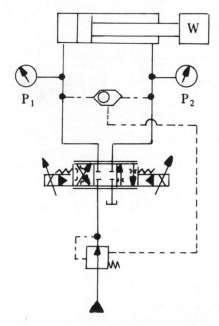

Fig. 5.6 Depending on load, when cylinder is extended, pressure at gage P1 can actually be less than at gage P2.

Here, depending on load size, when the cylinder extends, pressure *P1* can actually be *less* than *P2*. This will cause the shuttle valve to sense fluid pressure from the cylinder head end while it is extending: the meter-in pressure compensation that was originally intended is eliminated, see calculations for resistive loads. To avoid this problem, a 2-position directional valve can be used, Figure 5.7.

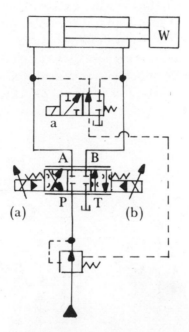

Fig. 5.7 To avoid inadvertent loss of pressure compensation, replace shuttle valve with 4-port, 2-position directional control valve.

By keeping the 2-position, directional control valve de-energized when the cylinder extends, only pressure in the line from port *A* is felt at the spring chamber of the pressure reducing element. When the cylinder retracts, after solenoid (a) on both of the directional valves have been energized. The spring chamber of the reducing element senses only pressure in the line from port *B*. This arrangement provides safe meter-in pressure compensation for both directions of travel of differential cylinders. This type of circuitry is useful when frictional load conditions allow the proportional valve to generate the necessary counterbalancing and braking forces without the assistance of counterbalance valving.

METER-OUT PRESSURE COMPENSATION

Meter-out, pressure compensated systems can be designed with conventional pressure reducing valves, or logic cartridge valves. These are discussed later in this chapter. Although meter-out circuits can prevent a suspended load from "pulling" an actuator ahead of pump supply, they are seldom used because they can cause damaging pressure intensification in the cylinder. Let us consider several possibilities.

In the circuit in Figure 5.8, two conventional pilot-operated pressure reducing valves reduce fluid pressure in ports A and B on

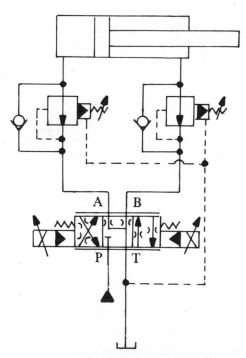

Fig. 5.8 Two conventional, pilot-operated, pressure reducing valves reduce pressure on proportional valve ports **A** *and* **B**.

the proportional valve. Actually, fluid pressure on the valve actuator ports is the *low* pressure setting of the pressure reducing valve pilot (about 150 psi) *plus* the residual pressure of the tank return line felt in the valve drain ports.

Note that the pressure differential in the valve is adjustable so that a desired flow can be established at a relatively low pressure drop. However, this circuit can generate excessive pressure in the cylinder head end.

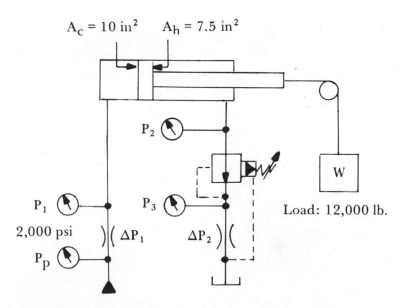

Fig. 5.9 Load and force analysis diagram of portions of circuit illustrated in Figure 5.8.

In the circuit illustrated in Figure 5.9, when the cylinder head end (7.5 in^2) is pressurized to 2000 psi, it could lift a load of 15,000 lb. If it is assumed that the cylinder is extending while lowering a load of only 12,000 lb, we must analyze the pressures in the cylinder to understand and appreciate the potential problem.

Assuming that the compensator is maintaining a 150-psi pressure drop on the outlet orifice, gage P_3 will read 150 psi, if tank port pressure is zero. With a 1:1 area ratio proportional valve and this cylinder:

$$Flow_1 / Flow_2 = \sqrt{\Delta P1} / \sqrt{\Delta P2}$$

$$10/7.5 = \sqrt{\Delta P1} / \sqrt{150}$$

$$\Delta P1 = [(10 \cdot \sqrt{150})/(7.5)]^2 = 266 \text{ psi}$$

With a pressure drop of 266 psi on the in-feeding orifice, gage P1 reads:

$$2000 - 266 = 1734 \text{ psi}$$

When the load is moving at constant velocity, there must also be a force balance. In this example, the load is pulling with a force of 12,000 lb, while, at the same time, the cylinder is pushing the load in the *same* direction with a force of 17,340 lb:

$$1734 \text{ psi} \cdot 10 \text{ in}^2 = 17,340 \text{ lb.}$$

The sum of these two forces (17,340 + 12,000) must be balanced by the backpressure which the compensator generates on the cylinder head end. Consequently, gage P2 will read:

$$29,340 \text{ lb}/7.5 \text{ in}^2 = 3912 \text{ psi}$$

For this 2000-psi system to operate safely, the cylinder would have to be rated at 5000 psi. As with conventional meter-out circuits, this problem becomes even more acute when larger rod cylinders are used.

METER-OUT LOAD COMPENSATION

The load locking, load compensating, meter-out sandwich plate is for use with pilot operated proportional valves. Although the valve incorporates three separate functions, only two can be achieved simultaneously, depending on circuit design.

The three functions are:

- leakfree load holding with a pilot operated check valve
- overcenter counterbalancing when ports A and/or B are connected to tank, and
- a meter-out pressure compensator if an orifice is placed downstream of the valve's A and/or B ports.

Although the valve provides control in both directions of actuator movement only one direction of motion is described here, for sake of simplicity.

In Figure 5.10, the proportional valve is mounted on top of the sandwich plate with ports A and B connected in series with ports A

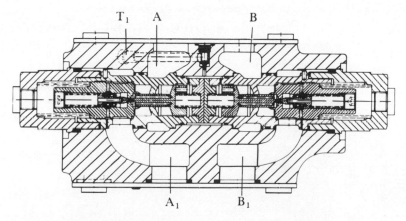

*Fig. 5.10 Proportional valve is mounted on top of sandwich plate. Valve ports **A** and **B** are connected in series with ports **A** and **B** in sandwich plate.*

and B in the sandwich plate. Also, port $T1$ in the sandwich plate is teed into the tank port passage of the proportional valve. Ports $A1$ and $B1$ are connected directly to the actuator working ports through the valve subplate.

When the proportional valve is in center position, either a fully closed center or a bleed center spool orificed to tank can be used. If an optional relief valve is used instead of orifice 10, however, Figure 5.11, a valve with a center spool position which is orificed to tank should be used.

Static Condition: Load Holding

When the proportional valve is centered, Figure 5.12, pressure in port passages A and B, between the proportional valve and the sandwich plate are vented to tank through orifice 10 and/or the centered proportional valve. This prevents any pressure build up in chambers $8a$ and $8b$, nulling all forces acting on pilot pistons $4a$ and $4b$, Figure 5.12. Pilot operated check valve poppets $1a$ and $1b$ lock the load in position.

Notice that load pressure in ports A and B passes through radial holes and passages $10a$ and $10b$ in poppets $2a$ and $2b$ to help springs $3a$ and $3b$ which have a pressure equivalent of about 60 psi. Thus, spring force plus load-induced pressure force hold poppets $1a$ and $1b$ in a leakfree, seated position.

Case 1: Load Holding with Meter-out Pressure Compensation

Referring to the schematic in Figure 5.11, and the cross-sectional drawing of Figure 5.12, when the proportional valve is shifted to

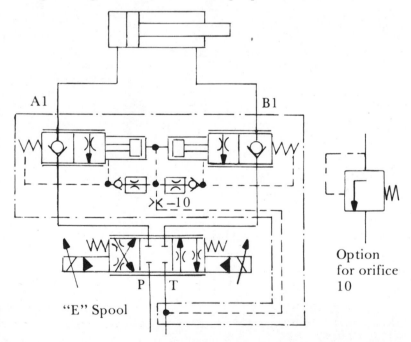

Fig. 5.11 When selecting an optional relief valve in place of orifice, specify spool which, when in center position, connects orifice to tank.

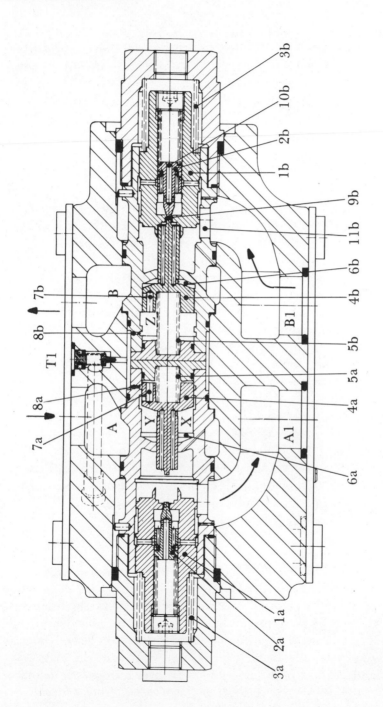

Fig. 5.12 Cross sectional drawing of load holding, meter-out compensator/counterbalance sandwich plate in operating mode.

extend the cylinder, pump output flows from port A to $A1$ in the free-flow direction of check valve $1a$. This line becomes pressurized because of the resistance to flow offered by the cylinder piston, Figure 5.11.

At this point, pilot flow is also established over pressure compensated flow control $7a$. A backpressure of about 175 psi is felt in chamber $8b$ and on pilot piston $4b$ because of the resistance to flow created by fixed orifice 10 or an optional relief valve. The optional ball type, fixed relief valve is set at 175 psi.

As pressure-induced force moves pilot piston $4b$ to the right, pilot operated poppet $2b$ also moves to the right, blocking the previously open connection between $B1$ and the spring chamber of spring $3b$. Fluid pressure at port B is now exposed to spring chamber $3b$ and to the annular area to the right side of pilot piston $4b$ via radial hole $6b$, axial holes $10b$ in pilot poppet $2b$, and the axial holes in the head end of pilot poppet $2b$.

Now, fluid pressure in port B acts on an effective area which consists of the full diameter of pilot piston $4b$ and helps the 60-psi equivalent force of spring $3b$. The sum of these two forces works against the constant force induced by the pressure of 175 psi created by orifice 10 on the opposite side of pilot piston $4b$.

The valve now modulates an orifice at radial holes $11b$, to maintain a constant pressure of 115 psi $(175 - 60)$ in port B. This pressure compensates the meter-out orifice passage B-to-T as created by the proportional valve. If a backpressure is felt in the tank port of the proportional valve, an equal amount of backpressure is felt at the outlet of orifice 10 or optional relief valve. This increases the opening forces acting on the pilot piston, thus maintaining the required 115-psi pressure drop on the B-to-T passage.

Case 2: Load Holding and
Overcenter Counterbalancing

As mentioned in Case 1, above, the pull of the load and the pushing force exerted by the pump both create an undesirably high pressure at the outlet of the actuator. In addition, if we consider the possible pressure intensification of differential-area cylinders, outlet pressure on the actuator can become dangerously high. If your calculations of actuator and load conditions show that pressures in the actuator become excessive, consider using the overcenter counterbalancing effect of the pressure compensator plate.

In Figure 5.13, a cylinder with nearly a 2:1 area ratio, is used to conterbalance an overcenter load condition. Notice that the shifted positions of the proportional valve spool provides no orificing in passages A-to-T or B-to-T. Consequently, the proportional valve can only meter-in oil to the actuator. In this circuit, a pressure compensated, meter-in sandwich plate is also used to provide constant speed with variable load conditions, as shown.

Because the proportional valve provides little or no backpressure in passage B, Figure 5.12, the only closing force acting on poppet $1b$ is 60 psi, which is the equivalent to the force exerted by spring $3b$. At first, as the proportional valve shifts to create an orificed passage P-to-A, a resistive load will generate high fluid pressure in port A of the sandwich plate.

If this high pressure fluid pressurizes the inlet side of the pressure compensated flow control $7a$, a 175-psi pilot pressure is created, as previously discussed. This 175-psi-pressure induced opening force can drive poppet $4b$ to a wide open position against the 60 psi equivalent force exerted by the spring.

As the load condition illustrated in Figure 5.13 approaches top dead center, resistance to flow, and therefore the pressure in line A-to-$A1$ decreases. As pressure at the inlet to the pressure compensated flow control drops below 175 psi (just before top dead center) pressure at the outlet must also decay. A modulation point is reached when pressure on pilot piston $4b$ drops to 60 psi. At this point, the spring force acting behind poppet $1b$ is enough to seat the poppet.

While the load is being lowered, the poppet creates an orifice at radial holes, $11b$, Figure 5.12, which resist flow sufficiently to prevent the load from overrunning. To keep the load moving while it is being lowered, a pressure induced force of only 60 psi is needed on the cylinder cap end.

PRESSURE COMPENSATION OPTIONS
USING LOGIC ELEMENTS

When installing pressure compensating proportional valves on a manifold, the designer has the option of using logic cartridge valves.

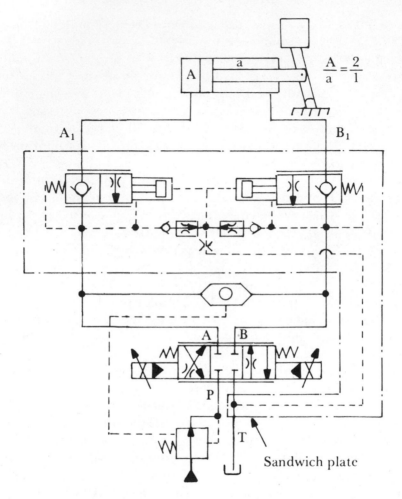

Fig. 5.13 Center position of proportional valve provides no orificing to passages A to T or B to T.

Virtually any pressure control function can be performed with a cartridge valve, and since logic cartridge valves were designed specifically for mounting in manifolds, they provide complete, compact units.

In review, a logic cartridge valve consists of a cartridge assembly and a cover. The cartridge assembly, which in turn, consists of a sleeve, a poppet, and biasing spring, is mounted in a manifold. The cover, which is mounted on top of the cartridge assembly, performs two functions: it holds the cartridge assembly in its bore and controls pilot oil flow on top of the cartridge assembly. Depending on poppet type and cover, pressure compensation for proportional valves can be accomplished with logic pressure relief elements, Figure 5.14 (*a*), and pressure reducing elements, Figure 5.14 (*b*).

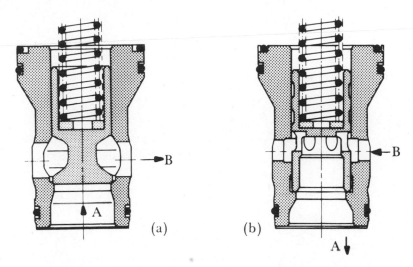

(a)   (b)

Fig. 5.14 Depending on poppet type and cover, pressure compensation for proportional valves can be achieved with logic pressure relief element (a) and pressure reducing element (b).

For meter-in, 2-way pressure compensation, a normally open pressure reducing element may be specified, Figure 5.15. The valve may be pressure compensated from ports *P-to-A* or *P-to-B*. The pressure drop maintained across the proportional valve can be either 75 or 120 psi. This is achieved by changing the biasing spring on top of the poppet.

For 2-way meter-in pressure compensation, when the designer wants an adjustable pressure drop across the proportional valve, a normally open, pressure reducing element is used. An adjustable relief valve is included in the cover, Figure 5.16. A light spring equivalent to a force exerted by a pressure of 45 psi is needed in the reducing element to hold it in its normally open position. The relief valve adjusts the pressure drop across the proportional valve, by adding an adjustable spring force to the load induced pressure force. This was discussed in detail on page 105.

For meter-out compensation, a normally open reducing element

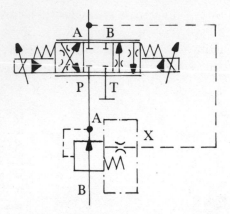

Fig. 5.15   For meter-in, 2-way pressure compensation, N.O. pressure reducing element (with appropriate cover) may be used.

with its respective cover can be placed in line A, for instance, Figure 5.17. This provides constant return flow from port A-to-T. The pressure drop maintained across the proportional valve can be either 75 or 120 psi. Again, this pressure is established by the bias spring on top of the element.

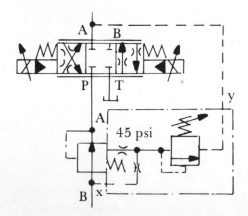

Fig. 5.16   Relief valve adjusts pressure drop across proportional valve by adding adjustable spring forces to load-induced pressure force.

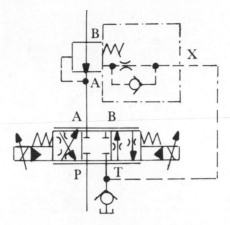

*Fig. 5.17 For meter-out pressure compensation, N.O. pressure reducing
element (with appropriate cover) can be mounted in line A
to provide constant return flow from A to T.*

For proper operation, a 75-psi check valve is needed in the tank
line to maintain backpressure on the spring side of the element.
This backpressure, acting through the reverse free flow check valve,
quickly provides the minimum necessary opening forces as well as
stable position.

For an adjustable pressure drop across the proportional valve, a
N.O. pressure reducing element is placed in line *A*, Figure 5.18. A
cover with an adjustable, direct-operated relief valve is used to
establish the pressure drop. Return flow from port *A*-to-*T* is constant.
A light bias spring equivalent to a pressure of about 45 psi, the element
open, while differential pressure is set on the pilot valve in the cover.

For 3-way bypass type pressure compensation, a normally closed
pressure relief element is used, Figure 5.19. This bias spring on top
of the element is equivalent to a pressure of about 75 to 120 psi. As
soon as inlet pressure equals spring value higher than outlet pressure,
the element opens to maintain a constant pressure drop across the
proportional valve, equivalent to the force exerted by the spring.

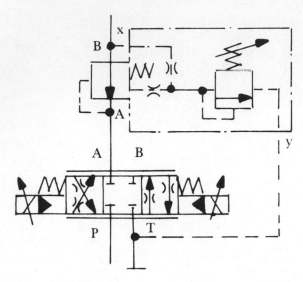

*Fig. 5.18   For adjustable pressure drop across proportional valve, N.O.
pressure reducing element is placed in line A.*

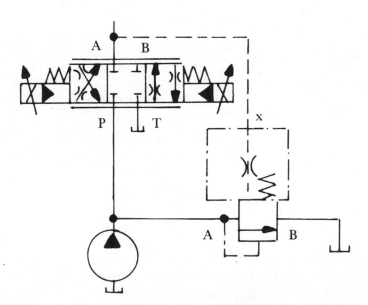

*Fig. 5.19   For 3-way, bypass type pressure compensation, NC pressure
relief valve is used.*

Meter-in, 3-way pressure compensation can be achieved from ports
P-to-*A* and *P*-to-*B*.

For maximum pressure protection, a direct operated relief valve
is available in the cover, Figure 5.20. If load induced pressure exceeds
the pressure requirements of the system, the direct operated relief

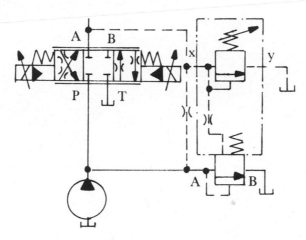

*Fig. 5.20 For maximum pressure protection, direct operated relief valve is
mounted in cover.*

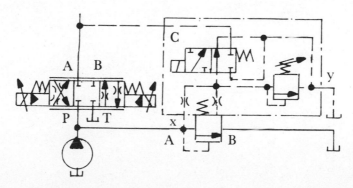

*Fig. 5.21 Three-way, pressure compensation with maximum pressure protection
can also incorporate unloading function.*

valve opens, limiting the maximum load induced pressure force sensed on top of the logic cartridge. Since the element incorporates a 45-psi spring, the element opens at a pressure level that is 45 psi *higher* than the setting of the direct operated relief valve.

Three-way pressure compensation with maximum protection can also incorporate an unloading function, Figure 5.21. The 2-position directional valve, *C*, is shown in its unloaded position. When the directional solenoid valve, *C*, is energized, 3-way pressure compensation is achievable. Although it was not previously stated, 3-way by-pass style flow regulation can also be accomplished with conventional relief valves or vented solenoid relief valves plumbed separately in the circuit.

REVIEW QUESTIONS

5.1   In what ways are electronic proportional valves unique?

5.2   What is load compensation in a valve? In a pump?

5.3   When would a valve need load compensation?

5.4   Discuss the basic requirements of load compensation.

5.5   What is the basic principle involved in any load sensing function?

5.6   Must special considerations be taken into account when dealing with proportional, 4-port valves? Explain.

5.7   Refer to Figure 5.1, what conditions must be satisfied under all modes of operation and load conditions?

5.8   Under what conditions should a designer consider a counterbalance or overcenter counterbalance valve between the cylinder and proportional valve? Explain.

5.9   When designing a circuit with a pressure compensated proportional valve, what must the designer be especially careful of? Why?

5.10   Describe a meter-in, pressure-compensator.

5.11   How many basic models of meter-in pressure compensators do you know of? What are they?

5.12   Is a pressure compensator a NO or NC device?

5.13   What is the spring setting of the compensator spring? Is this setting important? Why? Explain.

5.14   Referring to Figure 5.2, describe the action of the valve. Explain what happens in case of load-induced pressure fluctuations at either end of the valve spool.

5.15   When is meter-in pressure compensation generally used?

5.16   For any meter-in circuit, deceleration and overhung loads require the use of a counterbalance valve. True or false? Explain.

5.17   Referring to Figure 5.3, discuss the major features of a circuit using pressure compensated, meter-in sandwich plates.

5.18   The circuit in Figure 5.4 is similar to that in Figure 5.3 except for the use of overcenter counterbalance valves for variable conditions and a conventional pilot operated reducing valve for load compensation. Discuss.

5.19   Explain how the circuit in Figure 5.4 works and why.

5.20   Is it possible to obtain meter-in compensation without counterbalancing? Explain.

5.21   What is a resistive load?

5.22   What other types of loads are there?

5.23   Referring to Figure 5.5, explain the action of the shuttle valve.

5.24 Could such a shuttle valve arrangement be used with a differential area cylinder? Explain.

5.25 Discuss meter-out pressure compensation. Are these popular and frequently used? Explain why.

5.26 What is the potential problem with the circuit in Figure 5.8?

5.27 When would you use a load-locking, load compensating, meter-out sandwich plate? Why?

5.28 Explain in detail how the valve illustrated in Figures 5.10 and 5.12 works?

5.29 Discuss the load-holding with meter-out pressure compensation system and the load-holding and overcenter counterbalancing.

5.30 How else can pressure compensation be achieved?

5.31 What is a cartridge valve? Name several types.

5.32 Can cartridge valves be used to perform any pressure control function? Explain.

5.33 What elements does a cartridge valve consist of?

5.34 What functions does the cartridge valve cover perform?

5.35 Are cartridge type valves usually poppet style? Why? Explain.

5.36 Can cartridge valves be used for meter-in functions? How about meter-out functions? How?

5.37 Discuss whether (if so, how) you can obtain meter-in and meter-out pressure compensation.

Chapter 6

ELECTRONICS

To operate any proportional solenoid, two major electronic elements are needed: an amplifier and a power supply. The electronic amplifier, provides the driving current to the proportional solenoid and interfaces the control signal. The power supply converts 120 volts AC to 24 volts DC to drive the card, Figure 6.1

To distinguish between the various types of amplifiers, each will be described separately. The first one is referred to as a single force controlled solenoid. It is primarily used to control pilot operated proportional pressure relief and pressure reducing valves. Other uses include control of single-solenoid, two-position, pilot-operated proportional directional valves and proportional pumps and motors, all equipped with force solenoids. The single force controlled solenoid type amplifier has the simplest internal electronic circuitry and requires the least amount of wiring. Figure 6.2 is a functional block diagram of a single force controlled solenoid.

The designer should concern himself mostly with learning how to wire the card and how to use its adjustments. He should *not* become involved with the internal electronic circuitry. The diagram shows an input and output side where all external wiring takes place. Everything between the dashed two lines is already on the amplifier card.

Let us begin with the input side where power is supplied and where a command signal must be provided. When supplying power to the card, a specific polarity must be observed: the positive and negative leads from the power supply must be connected to the proper

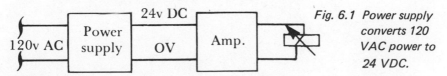

Fig. 6.1 Power supply converts 120 VAC power to 24 VDC.

terminals on the card. Should these leads be switched inadvertently, the card will obviously be unable to function properly.

To operate the valve, a low level command signal is provided on the input side. This is achieved by providing reference voltage terminals +9V, 0V, -9V, at the points where either a switch or potentiometer can be added. A potentiometer is usually preferred over a switch because its command signal is adjustable by simply turning a knob. At the output side, terminals are provided for wiring the amplifier card to the force solenoid. Also, the amplifier can control only one force solenoid at one time.

Internal Circuitry

Each block in the amplifier card diagram represents a specific function.

As power is supplied to the card [terminal 24 ac (+), terminal 18 ac (-)], the first block the power signal encounters is a voltage filter and regulator (1). Since the 24V DC signal is not necessarily consistently smooth or steady, the voltage is regulated to provide a fixed known level of voltage (designated by terminals 10 ac (+9V), 14 ac (0V), and 16 ac (-9V)) which are used for the rest of the circuit. Voltage regulation also provides temperature stability which helps keep the card accurate over a wide temperature range: as ambient temperature varies, the set point on the card remains constant.

From these known levels of voltage called reference voltage terminals, a potentiometer or switch can be wired, to provide the

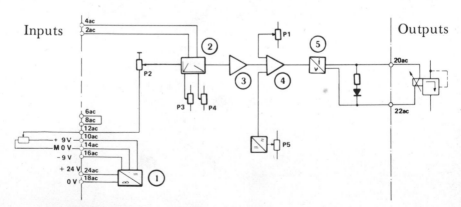

Fig. 6.2 Functional block diagram of single force controlled solenoid electronic amplifier.

required command signal. Figure 6.2 shows a potentiometer wired to the card. One leg of the potentiometer is connected to terminal 10 ac (+9V), the other to terminal 14 ac (0V). The wiper or output of the potentiometer is connected to terminal 12 ac. This arrangement provides a command signal range from 0 to +9V which can be adjusted by turning the potentiometer.

From 12 ac, the signal enters the second block, called a ramp generator (2), Figure 6.2, and comes out as a linear change with respect to time. This change is the generator's sole function.

The most important thing to remember about the ramp generator is the signal from the generator does not change in magnitude but that it reproduces the magnitude of the input at a given time rate. In other words, if the input goes from 0 to 100%, the output also goes from 0 to 100% but at a given time rate. Likewise, when the input goes from 100% to 0 the output goes from 100% to 0 at a given time rate. The ramp generator does not allow the output to move faster than the ramp adjustment setting. Ramp adjustments are discussed later.

From the ramp generator the output then proceeds to the next block, the matching amplifier (3), which is basically an amplification stage.

A summing amplifier (4), adds three signals: one signal from the dither oscillator which generates a modulating signal to minimize solenoid hysteresis; a signal from adjustment $P1$ which is a pilot or bias current setting; and the input signal.

The power or output amplifier (5) performs two basic functions: it amplifies the signal to the power level needed to drive the solenoid, and it provides current feedback to continue stable operation over changes in coil temperature and wiring losses.

AMPLIFIER ADJUSTMENTS

Every amplifier card has a number of adjustments that must be set for proper valve operation. These settings are usually called presets which are actually multiturn potentiometers.

Adjustment $P3$ and $P4$

As mentioned, the ramp generator can be adjusted. A single

force controlled solenoid has two ramp adjustments: one, designated *P3*, is for setting the ramp up time (the time it takes for an increasing signal), the other designated *P4*, is for setting the ramp down time (the time it takes for a decreasing signal). With two separate ramp adjustments, independent ramp times can be set. If, for example, the system needs a proportional pressure relief valve and slow pressure build-up with quick pump unloading by setting *P3* for a slow ramp *up time* the maximum pressure setting of the valve could take as long as 5 seconds. By setting *P4* for a fast ramp *down time*, the pump can be unloaded quickly. Likewise, *P3* can be set so pressure builds quickly and *P4* can be set to unload the pump slowly.

Acceleration, deceleration, and decompression can be achieved with the ramp generator. The given ramp time range for the amplifier card is from 0.1 second minimum to 5 seconds maximum. However, the maximum 5 second time limit is achievable only by providing a +9 volt signal to the input of the ramp generator. If the signal is less than +9V, the ramp time will not be achieved in 5 seconds as intended. The curve in Figure 6.3, illustrates this relationship.

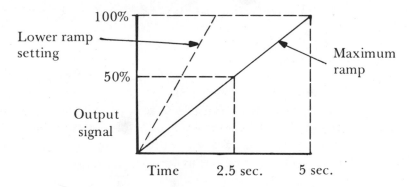

Fig. 6.3 *If signal is less than +9V, the ramp time will be less than 5 seconds.*

At 100% signal, the system achieves a maximum ramp time of 5 seconds. By changing the input signal to 50% and keeping the ramp time set at its maximum value, ramp time reduces to the change in signal. Note also that if no ramp time is required, terminals 4 ac and 2 ac, Figure 6.13, can be bridged by either a jumper wire or a switch. By jumping terminals 4 ac and 2 ac, the signal bypasses the ramp generator completely. This means that the output follows the input, or the valve will respond directly to the preset $P1$ and $P2$ or the remote potentiometer setting.

The force out of a proportional valve is linear only over a particular range. Proportional valves are designed to work over this linear range and below the nonlinear point on the curve becomes unpredictable. To overcome this problem, preset $P1$, Figure 6.4 (which is a bias current setting), adds a signal to boost the input signal to that useable

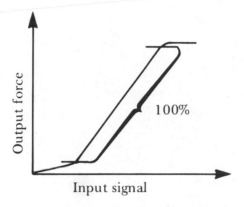

Fig. 6.4 At 100% signal, system achieves ramp time in 5 seconds.

linear portion of the curve. This allows the force output of the valve to work only over this useable linear range. Because this amplifier is not used exclusively for proportional pressure relief valves, and because this nonlinear point differs for other components that require the use of this amplifier card, $P1$ can adjust a bias current from 0 to 300 mA. To further demonstrate this point, a performance curve for a proportional pressure relief valve is given in Figure 6.5.

Figure 6.5 shows that a portion of the curve is called dead range or deadband. Dead range is the area where the setting of $P1$ is of utmost importance. For example, assume that $P1$ is set at zero and a potentiometer will provide the command signal. As the potentiometer is turned, the valve will *not* build pressure until the signal reaches about 180 mA. The signal must first travel through the dead range before the valve will respond. To boost the input signal to its proper starting point, $P1$ must be adjusted to provide a 180 mA signal. This way, when the operator first begins to turn a potentiometer, the valve immediately starts to build pressure. $P1$ would be set at about 180 mA only assuming the lowest set pressure related to flow is about 100 psi.

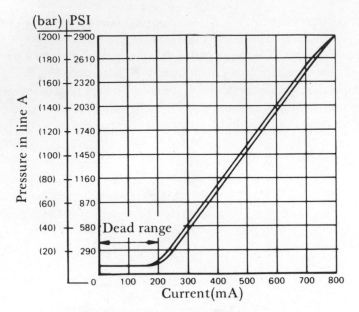

Fig. 6.5 *Performance curve for proportional relief valve.*

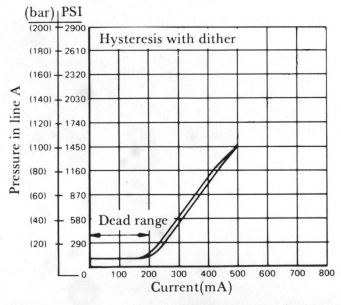

Fig. 6.6 *Performance curve for proportional relief valve with P2 adjusted to limit output current to approximately 490 ma.*

To set the maximum pressure of the valve, preset $P2$ must be used. $P2$ is a maximum current limitation adjustment, which is accessible on the front face of the card. $P2$ can be adjusted to limit the command signal, so if the full pressure range for the proportional pressure relief valve in Figure 6.6 is desired, and assuming $P1$ is set for 180 mA, $P2$ would have to be set for its maximum value of 800 mA. If a lower valve pressure setting is required, $P2$ can be adjusted accordingly. Also, since current is directly related to the pressure setting of the valve, pressure in the system can be adjusted by presets $P1$ and $P2$ while reading the pressure gauge at the same time. A 180 mA signal corresponds directly to approximately 100 psi on the curve; similarly, an 800 mA signal corresponds directly to 2900 psi.

The same holds true for any other desired limits. For example, if an application calls for a maximum pressure of 2500 psi and the minimum pressure setting of the valve related to flow is 150 psi, the lower and upper limits can once again be established by setting $P1$ and $P2$. $P1$ would be adjusted to set the lower 150 psi limit, and $P2$ the upper 2500 psi limit. Also with a potentiometer, the valve could provide variable adjustment between 150 to 2500 psi without exceeding the upper limit.

Another factor to consider when adjusting $P1$ and $P2$ is that $P1$ *adds* to the setting of $P2$. If $P1$ is set at zero and $P2$ is adjusted for some *maximum* pressure, then $P1$ is adjusted to its *minimum* pressure setting. Because the pressure setting of the valve will increase

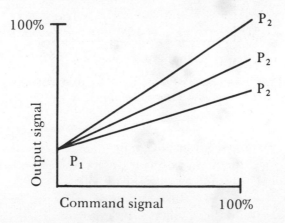

Fig. 6.7 Pressure setting of valve will increase to setting of P1.

to the setting of $P1$, Figure 6.7, it is necessary to first establish $P1$, then set $P2$ for its maximum value. This is done by holding the command signal at zero and setting $P1$ for the required bias current. $P2$ can then be set by setting the command signal at 100% and setting the maximum current as desired. The following curves will illustrate these conditions in more detail.

When $P1$ is set for its minimum bias current, $P2$ can be increased or decreased to the maximum desired current without changing the bias current setting. Relating this to the proportional pressure relief valve, and knowing that $P2$ sets the maximum current level or maximum pressure, maximum pressure can be increased or decreased without changing the minimum pressure setting of the valve.

Changing the setting of $P1$ when $P2$ is already set for its maximum pressure, changes the entire setting of the valve. When $P1$ (the bias current) is increased, Figure 6.8, the range or span remains the

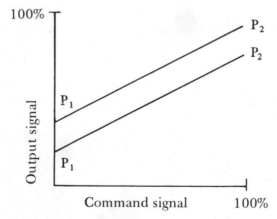

Fig. 6.8 When bias current P1 is increased, range remains the same but at higher level lower and upper limits.

same, but at increased lower and upper limits. Pressure, for example would rise from a minimum of 100 psi to a *minimum* of 200 psi. If maximum valve pressure was originally set for 2500 psi, it would also increase 100 psi to a final setting of 2600 psi. Preset $P5$ is the adjustment for the dither oscillator mentioned earlier which is preset at the factory, and needs no adjustment.

HYSTERESIS

In proportional valves, hysteresis is caused by friction, as in a solenoid, spool, or poppet. Hysteresis in a proportional valve can be defined as the *difference* between the increasing input signal to the valve and the decreasing signal. Relating this to the spool position of a proportional directional valve, hysteresis can be defined as the difference (the error) in spool position as the spool approaches the set point from two opposite directions, Figure 6.9.

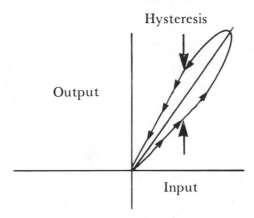

Fig. 6.9 Hysteresis is the difference in spool position as spool approaches set point from two opposite directions.

Hysteresis (error) in proportional valves is known as a percentage of error and in most good proportional valves it should not exceed 3%. Knowing the hysteresis of a valve enables the designer to establish a direct relationship relating to the accuracy of the valve. Thus, for example, if the hysteresis of a particular proportional valve is less than 3%, it can be said that its accuracy is better than 3% error.

The repeatability of a proportional valve is another term often used when describing the valve's dynamic characteristics. Repeatability is a measure of *exactness* with which motion or position can be duplicated. Exactness can also be defined as the error in output when the spool approaches the set point in the *same* direction. In other words, if the operator shifts a proportional directional valve twice in the *same* direction, the valve will not return to the exact same position both times. Repeatability is often one half of hysteresis

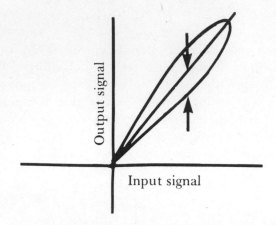

Fig. 6.10 Repeatability is usually half of hysteresis, but is never worse than hysteresis.

and will never be worse than hysteresis, Figure 6.10.

To help reduce hysteresis it was mentioned earlier in the internal circuitry section that the dither oscillator provides a modulating signal to minimize valve hysteresis. The following curves of a proportional pressure relief valve demonstrate the difference between a

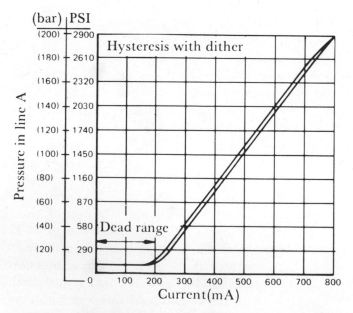

Fig. 6.11 (a) With dither signal, difference between increasing and decreasing pressure is about $\pm$ 2.5%.

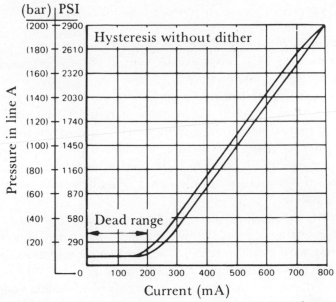

Fig. 6.11 (b) Without dither signal, difference is about ± 4.5%.

valve with hysteresis and one without.

Although this dither feature is built-in all electronic amplifiers the curves clearly show that there can be a substantial difference in pressure settings when a dither signal is excluded. With a dither signal, Figure 6.11 (a), the difference between an increasing and decreasing pressure setting is about ± 2.5%. Without the dither

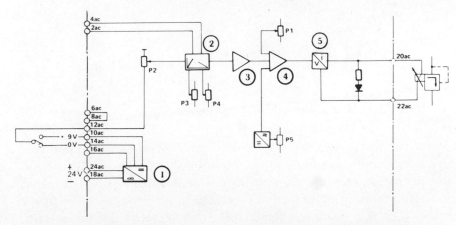

Fig. 6.12 For simple on-off control, single-pole, double-throw switch provides required command signal.

signal, Figure 6.11 *(b)*, the difference is about $\pm$ 4.5%.

Wiring Pattern for Single Force Controlled Solenoid Cards

The following diagrams illustrate typical wiring patterns for proportional pressure control valves.

For simple on-off control, a single-pole-double-throw switch can provide the required command signal, Figure 6.12. Since the switch is shown in the off position the minimum pressure setting of the valve depends on $P1$. When the switch is made to provide a command signal, the valve builds pressure to its maximum limit dependent on the setting of $P2$, and the time it takes depends on the setting of $P3$ for ramp up time.

When the switch is broken, the valve decreases in pressure dependent on the setting of $P4$ which is the ramp down time. This is one of the simplest wiring methods for achieving a single maximum pressure setting when setting time must be controlled.

The most common wiring pattern to provide a variable command signal quickly and conveniently, is with a potentiometer, Figure 6.13.

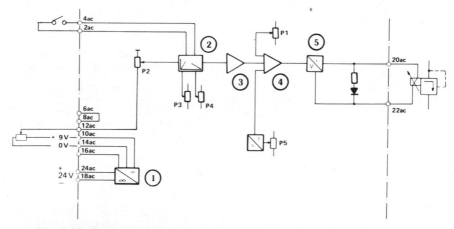

Fig. 6.13 Potentiometer is common wiring method to provide command signal.

As mentioned earlier, a potentiometer provides a variable command signal by simply turning a knob. When excited by DC or AC voltage, the potentiometer provides a proportional voltage versus displacement relationship. The potentiometer output signal is linear to the rotation of the potentiometer. As shown in Figure 6.13, the external potentiometer gives the operator full control of the pressure range of the valve as long as *P2* is set for its maximum current. Since *P2* is connected *in series* with the potentiometer, *P2* can determine the maximum pressure limit, whereas the potentiometer can be adjusted to whatever pressure is required not to exceed the setting of *P2*.

Also, if the ramp generator is turned off, the operator can control the speed at which pressure increases or decreases by the rate at which he turns the potentiometer knob. When the ramp generator is ON, it does not matter how fast the potentiometer is turned because the ramp generator will allow the signal to reproduce only at the time set. Obviously, one can turn the potentiometer slower than the ramp setting. This arrangement is very useful in test situations because the potentiometer can be mounted remotely from the test with the option of controlling pressure with or without the ramp time. Also note that the external potentiometer should have a resistance no lower than 500Ω nor higher than 5000Ω.

To cycle back and forth from several different pressure settings, and to maintain these settings without having to hand-adjust a potentiometer each time a different pressure setting is required, the following wiring method can be used.

The diagram in Figure 6.14 shows four potentiometers wired in parallel and four switches in series. Each potentiometer or preset can be adjusted independently to provide a particular command signal which, in turn, can generate four separate pressure settings. Also note that because the switches are wired in series, an order of priority is established.

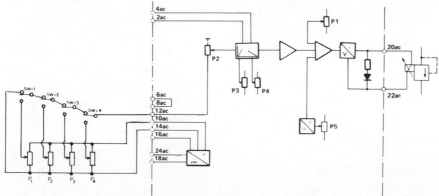

Fig. 6.14 Each of four potentiometers in circuit has a resistance of 5000 ohms.

If switch SW-1 is made to provide a signal from $P1$ and switch SW-2 is then made to provide a signal from $P2$, switch SW-2 will have priority and the signal from SW-1 will have nowhere to go because SW-2 has broken that part of the circuit. Switch SW-4 has the highest priority. If SW-4 is made to provide a signal from $P4$, it does not matter what position any of the other switches are in. The signal from $P4$ will always proceed first.

Although four potentiometers are shown wired in parallel, up to ten potentiometers may be wired in parallel with ten separate switches wired in series to provide various command signals. The important thing to remember when wiring potentiometers in parallel, is that the minimum resistance to the low level signal must never be lower than 500Ω.

Kirchoff's Law states that for resistors wired in parallel, the total resistance is the reciprocal of the combined resistances, which equals the sum of the reciprocals of each individual branch. Thus,

$$1/R_T = 1/R_1 + 1/R_2 + 1/R_3 + \ldots$$

If, for example, each of the four potentiometers in Figure 6.14 has a resistance of 5000Ω, the total resistance can be calculated from Kirchoff's Law:

$$1/R_T = 1/5000\Omega_1 + 1/5000\Omega_2 + 1/5000\Omega_3 + 1/5000\Omega_4$$

$$1/R_T = 4/5000\Omega$$

$$R_T = 5000\Omega/4 = 1250\Omega$$

The four potentiometers in parallel have a total resistance of 1250Ω, well above the required 500Ω minimum. Obviously, if more than ten potentiometers are wired in parallel, the minimum of 500Ω will be exceeded.

$$1/R_T = 1/5000\Omega_1 + 1/5000\Omega_2 + 1/5000\Omega_3 + \ldots + 1/5000\Omega_{11}$$

$$1/R_T = 11/5000\Omega$$

$$R_T = 455\Omega$$

Ten potentiometers wired in parallel each with a resistance of 5000Ω would equal 500Ω, the exact minimum requirement. These three wiring examples illustrate some of the possible methods that can be applied.

Electronic Amplifier Dual Force Controlled Solenoids

The amplifier card shown in Figure 6.15 *(a)* is used exclusively to control pilot operated proportional directional control valves and electronically controlled hydrostatic transmissions.

Basically, this amplifier operates like the single force controlled solenoid except that it is built to provide a signal to *two* proportional force solenoids and has some added internal circuitry to provide additional application uses.

Referring to Figure 6.15 *(b)* everything between the dashed lines is already on the card. External wiring of switches, potentiometers, and solenoids is done at the numbered terminals.

Fig. 6.15 (a) Photograph of amplifier card to control pilot operated proportional directional control valves.

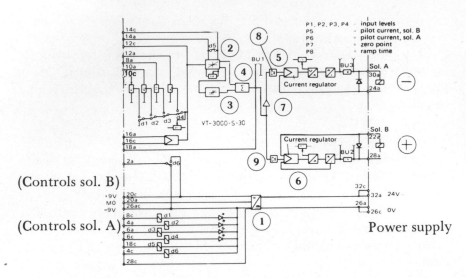

Fig. 6.15 (b) External wiring of switches, potentiometers, and solenoids for amplifier card in Figure 6.15 (a).

Internal Circuitry

Power of +24 volts DC must be supplied to the card [terminal 32 a (+), terminal 26 a (-)] and the first block it encounters is voltage regulator and filter (1). This is the same voltage regulation block on the single force controlled solenoid card, Figure 6.2. It provides a smooth steady, known level of voltage designated by terminals 20 c for +9 volts, 20 a for 0 volts, and 26 ac for -9 volts. To provide a signal to the next block, which is the ramp generator (2), the added presets $P1$, $P2$, $P3$ and $P4$ and relays $d1$, $d2$, $d3$ and $d4$ must be used. For sake of easy explanation, these presets and relays will be discussed later. At this point, assume that a signal is provided from the $\pm$ 9 volt terminals through any one of the four presets $P1$ to $P4$ to the input of the ramp generator.

As with the single force controlled solenoid, the signal enters the ramp generator and exits as a linear change with respect to time. The major difference is that instead of having separate ramp times for up and down, the generator only has one ramp time for setting the speed with which the spool opens and closes. This then relates directly to accelerating and decelerating times for load. Since the signal can be controlled, the valve is controlled as to how fast or slow and how far to open or close the valve, and smooth stopping and starting of loads can be achieved. These conditions were discussed earlier in the book as was typical acceleration and deceleration curves.

The output of the ramp generator then proceeds to the function generator block (3), Figure 6.15 *(b)*. The function generator is an added function which compensates for deadband in proportional directional valves. The primary causes for deadband in proportional directional valves are:

1. Solenoid friction.
2. Spring bias: springs which hold proportional spools in their center position are under tension and these forces must be overcome.
3. Spool overlap, which as stated previously, is precisely 11% for all proportional spools.

To better understand what the function generator does, assume this condition. If there was no function generator in the amplifier circuitry and a proportional directional was shifted in both directions, it would take a substantial amount of input signal before the valve would produce an output flow in either direction, Figure 6.16.

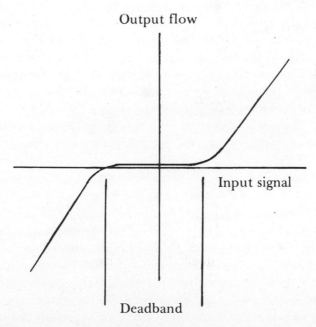

Fig. 6.16 *Substantial input signal is needed to produce output flow.*

A very large signal is needed to overcome solenoid friction, spring bias, and spool overlap (11%), which together create a significant amount of deadband around zero.

Thus, if a potentiometer is used to provide the command signal and an operator turns the potentiometer knob, nothing will happen until the potentiometer generates a large enough signal to overcome this amount of deadband. To overcome this undesirable condition, the function generator compensates for this deadband by abruptly jumping the input signal to the linear portion of the curve, Figure 6.17.

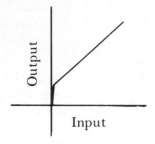

Fig. 6.17 To bypass unwanted condition, function generator compensates for deadband by abruptly jumping input signal.

The rest of the card operates basically like the single force controlled solenoid in which block 4 is the summing unit for adding the signal from the ramp generator and function generator. Blocks 5 and 6 are the power amplifiers for solenoids A and B, Figure 6.15 *(b)*.

Note that unlike the case in a single force controlled solenoid, negative and positive voltage (0 to $\pm$ 9V) must be provided to shift a 3-position 4-port proportional valve in both directions. In other words, a *negative* voltage command signal is needed to operate solenoid A, a *positive* voltage command signal for solenoid B. Inverter stage (7) and diodes (8) and (9) provide positive and negative voltage direction control.

1. Diode 8 allows positive voltage through
2. Diode 9 allows positive voltage through
3. Inverter stage A changes negative voltage to positive.

Thus, when a negative command signal is present, diode (9) rejects it because it only allows a positive signal through in the direction of the arrow. The inverter changes negative voltage to positive, allowing it to pass through diode (8) to operate solenoid A. Likewise, when there is a positive command signal, the inverter changes positive voltage to negative, so that diode (8) again rejects it and the signal proceeds through diode (9) to operate solenoid B.

Note also, that no dither oscillator is needed for a dual force controlled solenoid amplifier because the card operates according to a method called pulse width modulation.

Presets and Relays

The lower left hand segment of Figure 6.18 shows a section of relays indicated by a series of rectangles ($\boxed{/}$), numbered from $d1$ through $d6$. Also shown are the contacts for each relay (also numbered $d1$ through $d6$) in various locations throughout the amplifier.

To energize any one relay, 24 volts DC must be supplied to terminals 8 c, 4 a, 6 c, 18 c and 4 c. This can be done by coming directly off terminal 28 c (terminal 28 c is a 24 volt DC terminal) to a switch (or switches) and connecting the output of the switch to the particular relay terminal, Figure 6.18.

A light emitting diode (LED) connected in series with each relay $d1$ through $d4$ indicates when the relay is energized. These LEDs are numbered $d1$ through $d4$ and are located on the front face of the card. Above the four relays in Figure 6.18 are four more rectangles which represent four adjustable presets (accessible on the front of the card), numbered $P1$ through $P4$. These presets limit the $\pm$ 9V signal. Also accessible on the front of the card is $P8$ which sets the ramp time.

Presets $P1$ — $P4$

To generate a command signal, any of the relays, $d1$ through $d4$, must be energized. Figure 6.18 shows that the contacts of relays $d1$ through $d4$ are shown connected in series just below each preset $P1$ through $P4$. As was discussed with respect to the single force controlled solenoid, when switches or contacts are wired in series, the one with the highest priority (when activated) will always allow the signal to pass through *regardless* whether or not other contacts are being activated.

In this case, preset $P4$ has the highest priority. When relay $d4$ is energized, LED $d4$ comes on, and contact $d4$ breaks from the series wiring sequence and connects with the output of preset $P4$. This allows the signal to proceed to the input of the ramp generator and then through the rest of the card to its intended solenoid. When relay $d4$ is de-energized, the LED goes off and the signal drops to zero. Likewise, when relay $d3$ is energized, LED $d3$ comes on and

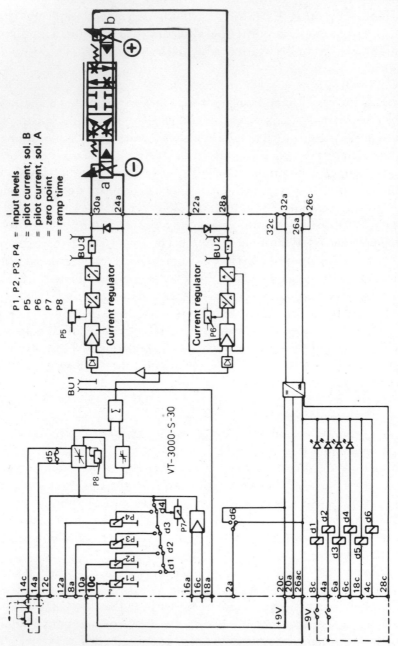

P1, P2, P3, P4 = input levels
P5 = pilot current, sol. B
P6 = pilot current, sol. A
P7 = zero point
P8 = ramp time

Fig. 6.18 To energize any one relay, 24 VDC must be supplied to several terminals.

contact $d3$ connects to the output of $P3$ allowing the signal to proceed through the card to its intended solenoid. The priority chain continues respectively with preset $P2$ corresponding to relay $d2$ and LED $d2$. Preset $P1$ has the lowest priority and corresponds to relay $d1$ and LED $d1$.

The diagram in Figure 6.18, represents a simple wiring schematic which shows how these presets and relays work in relation to a 3-position 4-port proportional directional control valve. First, note that a wire is connected from $20\ c$ to $10\ c$ to provide +9 volts; another wire is connected from $26\ ac$ to $10\ a$ to provide -9 volts. (Remember positive voltage is required to operate solenoid B and, negative voltage to operate solenoid A).

When relay $d1$ is energized, contact $d1$ pulls in, supplying a signal through the card to solenoid B. This shifts the proportional spool from its center position to a distance set by $P1$. In this case $P1$ can be said to be set to provide +9 volts, allowing the spool to travel its full stroke in one direction. De-energizing relay $d1$ shifts the spool back to its center position.

When relay $d2$ is energized contact $d2$ pulls in, sending a signal to solenoid A to shift the spool in the opposite direction, a distance set by $P2$. The spool again returns to center position when relay $d2$ is de-energized. Remember also that presets $P1$ and $P2$ can be adjusted to limit the $\pm$ 9V value and thus limit spool stroke. If other settings of the valve are required, presets $P3$ and $P4$ can be used. Typical wiring patterns showing all the presets in use will be discussed later.

Ramp Adjustment — P8

The dual force controlled solenoid amplifier card has only one ramp setting, $P8$, which is usually set within a range from 0.03 to 5 seconds. If necessary, the setting of $P8$ controls spool shifting (open and close) time. In the example given, if ramp time is set for its maximum of 5 seconds and relay $d1$ is energized, the signal will increase from 0 to 100% in 5 seconds shifting the spool from its center position to its set point in 5 seconds. Likewise, when $d1$ is de-energized, the signal will decrease from 100% to 0 in 5 seconds and the spool will shift from its set point back to center position in 5 seconds.

If no ramp time is required, relay $d5$ can be energized to close contacts $d5$ by bypassing the ramp generator completely; or terminals

14c and 14a can be bridged with either a switch or jumper wire. If an external time potentiometer is needed, it can be connected to terminals 14a and 14c, but the card potentiometer must be adjusted to its *maximum* ramp time, to minimize the possibility that the external potentiometer might exceed a time longer than that of the potentiometer on the card.

If, for example, the ramp setting on the card is 50% and the external time potentiometer is set at its maximum time, the signal will take the path of least resistance and the desired time will not be achieved. It is therefore necessary to keep the ramp setting at its maximum time when using an external time potentiometer. Also, if a preset is changed from a 100% signal to a lower value, and ramp time is held constant, ramp time automatically decreases. When selecting an external time potentiometer the rating of the potentiometer should be 500KΩ.

Preset 5, 6 and 7

Presets 5 and 6 are bias current settings and as with the single force controlled solenoid, they add a signal to boost the input signal to that useable linear portion of the curve. Unlike the single force controlled solenoid, these presets are set at the factory and require no adjustment. Preset $P7$ is a zero point, is also set at the factory, and needs no adjustment. These three presets are not accessible on the front face of the card.

Changeover Contact d6

By energizing relay $d6$ the changeover contact can be used to provide negative or positive voltage at any of the presets $P1$ through $P4$. A typical wiring diagram of this contact is shown in Figure 6.24.

For control systems with analog outputs, terminals 6a and 6c should be used. Terminal 12c can be used for an input for electrical joysticks.

Wiring Diagrams for
Dual Force Controlled Solenoid Amplifier Cards

One of the simplest ways to achieve bi-directional movement for a proportional directional valve and to establish various spool positions quickly and conveniently, is with an external potentiometer.

By connecting the potentiometer to 20c for +9 volts, 26ac for

-9 volts, and the wiper to any one of the preset terminals (in this case it is 10c for *P1*) the operator can control the valve spool in both directions. *P1* would determine the maximum spool travel in both directions, Figure 6.19.

Two Speeds Forward, Two Speeds Reverse

Many applications require a proportional directional valve and cylinder to accelerate a load to a constant velocity, then decelerate it to a slower or creep speed, and finally reverse the process to repeat the cycle. A single force controlled solenoid card and a few added electrical devices make this possible, see Figure 6.20.

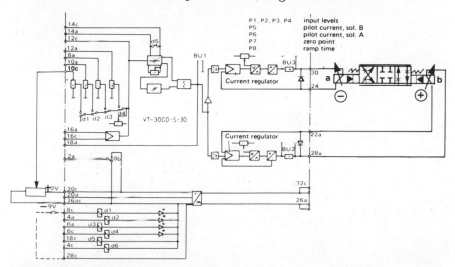

Fig. 6.19 Input P1 determines maximum valve spool travel in both directions.

Knowing that the card has four adjustable presets and two different speeds are required in each direction, a sequential order must be established when selecting the presets.

If the application calls for an *automatic* cycle, this can be accomplished by using external latching relays and establishing the relay logic accordingly. This can be seen more clearly in Figure 6.20.

First, two presets are wired to positive voltage; the other two for negative voltage, thus providing two forward and two reverse speeds when selected. Second, a typical relay logic diagram has been devised with four limit switches to provide the necessary logic to the amplifier card.

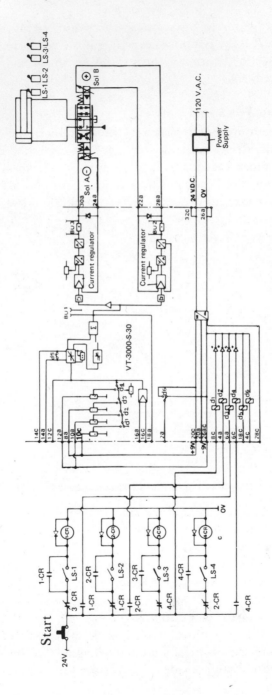

Fig. 6.20 Automatic cycle can be attained with external latching relays.

Starting with the relay logic diagram, Figure 6.21, when the operator depresses the start button, Figure 6.20, (assuming the cylinder is sitting directly on top of the first limit switch, *LS*-1) relay 1-*CR* will latch in, closing contact 1-*CR* and energizing internal relay $d3$.

Contact $d3$ closes allowing the cylinder to extend and accelerate to a speed set by preset *P3*. The cylinder extends until it closes limit switch *LS*-3 which in turn energizes relay 3-*CR* and opens normally closed contact 3-*CR*.

Relay 1-*CR* is de-energized, allowing the cylinder to decelerate to a slow extend speed, since the only relay on the card now energized is $d1$, with its preset *P1* set for some minimal value. Once the cylinder reaches limit switch *LS*-4, Figure 6.21, relay 4-*CR* is energized, closing contact 4-*CR*, energizing relay $d4$ and allowing the cylinder to accelerate as it retracts to a speed set at preset *P4*. The cylinder continues to retract until it closes limit switch *LS*-2, which energizes relay 2-*CR* and drops out relay 4-*CR*. At the same time, relay $d2$ is energized, decelerating the load to a creep speed at a minimal setting of *P2*. When the cylinder reaches limit switch *LS*-1, the cycle repeats.

Another important characteristic of the cycle is the ramp setting. Since there is only one ramp setting, all acceleration and deceleration values are the same, see velocity vs. time graph, Figure 6.21.

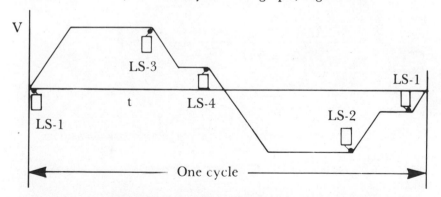

Fig. 6.21 In velocity vs. time diagram, all acceleration and deceleration values are the same.

When setting the ramp time for an automatic cycle, some fine tuning is generally needed to achieve smooth acceleration and deceleration.

Three Speeds Forward, One Reverse

Another typical wiring method often used in machine tool applications is shown in Figure 6.22. With three of the presets wired for +9 volts, three different speeds can be set during extend; the other preset wired for -9 volts can provide one speed for retract.

A typical velocity vs. time graph for this example is in Figure 6.23. By setting one of the presets for a high value, a fast or rapid traverse speed can be achieved. By setting another preset for a lower value, a feed speed can be achieved and by setting the last preset to a minimal value, a very slow cutting speed can be achieved. Once the cylinder reaches the end of its stroke, it retracts quickly to start the cycle again. The ramp is fine-tuned to achieve smooth speed change and reversal.

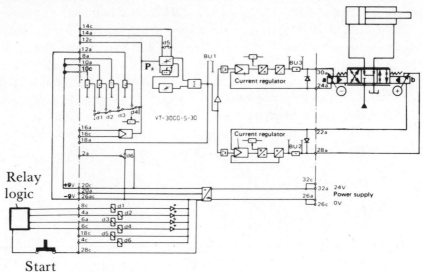

Fig. 6.22 Typical wiring method used in machine tool applications.

Four Speeds Forward, Four Speeds Reverse
Using Changeover Contacts

Figure 6.24 shows that contact $d6$ is tied directly to the +9 and -9 voltage lines and incorporates a separate terminal so it can be wired to any one of the preset terminals. By energizing relay $d6$, the contact switches from negative to positive voltage and back again when de-energized.

The example illustrates a 4-port proportional directional valve

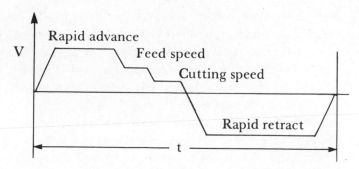

Fig. 6.23 Velocity vs. time graph for circuit in Figure 6.22.

controlling a hydraulic motor. The card is wired to all four presets from the changeover contacts.

It should be obvious that if each preset is adjusted for a different value, four different speeds can be obtained in either direction, since there would be a different spool position at each preset value. The direction can be changed with the changeover contacts.

These examples are but a small fraction of possible wiring methods which can be used with 4-port proportional directional valves. The wiring method used will depend on the application and the type of cycle to be accomplished.

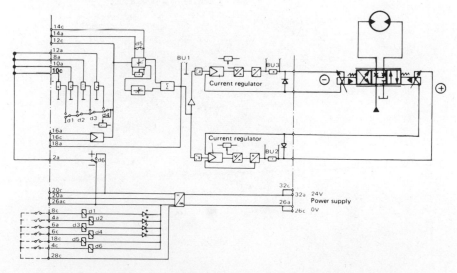

Fig. 6.24 Diagram for 4 speeds forward and 4 speeds reverse, using change-over contacts.

The dual force controlled solenoid (with multiple ramps) is similar to the dual force controlled solenoid amplifier card, Figure 6.25 *(a)*, except that it has five ramp settings (instead of one) all of which are accessible on the front face of the card. Figure 6.25 *(b)* shows that another board of relays has been added as well as an additional board of ramp adjustments.

The added relays are in parallel with the relays needed for $P1$ through $P4$. Anytime relays $d1$ are energized, contacts $d1$ pull in, which means that the set point depends on $P1$ and the ramp time on ramp setting $P11$. If relays $d2$ are energized, contacts $d2$ pull in, and the set point depends on $P2$ and ramp time on the ramp setting $P12$.

Thus, the priority chain remains the same for the ramp adjustments as it does for $P1$ through $P4$. Whenever relays $d4$ are energized, preset $P4$ and ramp setting $P14$ will have the highest priority.

Since each preset $P1$ through $P4$ can be set for different ramp time, acceleration and deceleration rates can be established separately for each preset. The ramp setting on $P10$ establishes the time required for the valve to center when all presets are deenergized.

Fig. 6.25 (a) Photograph of dual force controlled solenoid amplifier card.

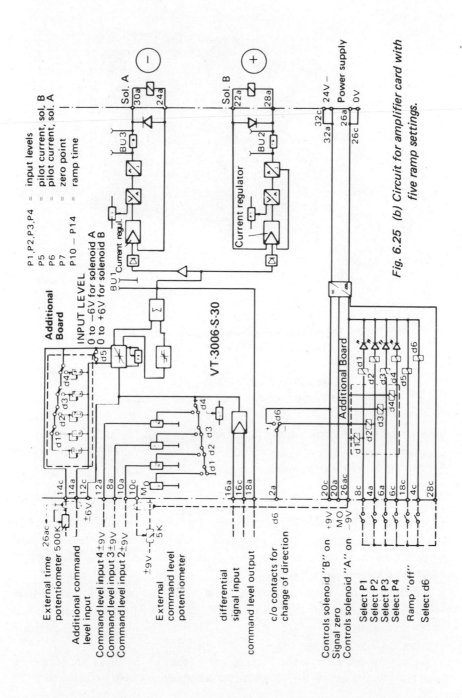

Fig. 6.25 (b) Circuit for amplifier card with five ramp settings.

*Fig. 6.26 (a) Photograph of single stroke controlled solenoid
electronic amplifier cards.*

ELECTRONIC AMPLIFIERS
SINGLE STROKE CONTROLLED SOLENOIDS

These electronic amplifiers, Figure 6.26 *(a)*, are used to control
pressure, directional and flow control valves with an LVDT mounted
on one proportional solenoid.

Internal Circuitry

Although the block diagram is the same for all single stroke
controlled solenoids, there are some minor differences (not shown)
between each card type. It is, therefore, important to use the correct
card for a particular valve.

Basically, there are two major differences between a force controlled solenoid card and a stroke controlled solenoid card. The latter has
cable break detection and some additional circuitry to compensate
for the feedback on the valves, Figure 6.26 *(b)*.

Starting with the added circuitry for feedback, there are four new
blocks. A proportional, integral, differential regulator, (P.I.D.)

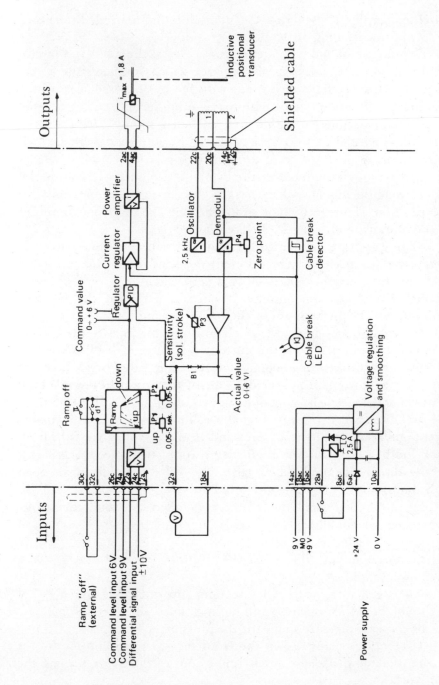

Fig. 6.26 (b) Cable break detection and additional circuitry compensate for feedback on valves.

matching amplifier, oscillator, and demodulator are needed to ensure the proper positioning of the orifice or spool.

The output of the ramp generator serves as the input to the proportional, integral, differential regulator which compares the actual position. This comparison is made possible by the oscillator, which is actually a separate circuit within the amplifier, generating a signal at a definite frequency within desired limits to the LVDT on the particular valve. The LVDT then sends to the amplifier a signal which corresponds to the position of the orifice or spool.

The demodulator receives the signal which recovers the intelligence from the signal and delivers a voltage signal proportional to the position of the orifice or spool through the matching amplifier to the proportional, integral, differential regulator.

At this point, a comparison is made in the proportional, integral, differential (as previously mentioned) resulting in a corrected signal back to the solenoid, maintaining a very accurate orifice or spool setting. The matching amplifier (which is preset at the factory) limits the stroke of the spool or the orifice setting.

The cable break detector, which monitors the lines to the LVDT, is connected internally to the current regulator. If feedback is lost because of a broken or unconnected wire, the cable break detector switches the current regulator off, cutting off power to the solenoid. Simultaneously, a light emitting diode (LED) on the front plate of the amplifier will turn on to signal a cable break. In cases involving direct-operated, directional valves and flow control valves with feedback, both valves will fail in *closed* position. When feedback is lost to the direct operated relief valve, the valve will fail in an open position to prevent any pressure build up.

Like the single and dual force controlled solenoid, the stroke controlled solenoid has:
1. Voltage regulator and filter
2. Current regulator to stabilize the output
3. Power amplifier
4. Ramp generator with separate ramp up and down times, to set spool or orifice opening and closing times or to set the increase or decrease pressure times for the pressure controls.

The only adjustments on the front face of the amplifier are for setting the ramp up and down times and a switch for turning the ramp off.

Feedback Connector and Wiring

It is of utmost importance that the ground terminal (symbol ⏚) on the feedback connector be wired properly.

Here is the wiring sequence: pin number 1 on the plug connects to terminal 20c; pin number 2 on the plug connects to terminal 14c on the card; and pin (⏚) on the plug connects to terminal 22c.

Electronic amplifier, dual stroke controlled solenoid, Figure 6.27 *(a)*, controls direct-operated proportional directional valves equipped

Fig. 6.27 (a) Photograph of dual stroke controlled solenoid electronic amplifier card.

with two proportional solenoids and a LVDT for positional feedback. The amplifier is the same as the dual force controlled solenoid, except that it has added circuitry for cable break detection and feedback as discussed for the single stroke controlled solenoid amplifier.

The presets and relays can be used the same way they are with the dual force controlled solenoid, Figure 6.27 *(b)*. The only significant difference is that, unlike the dual force controlled solenoid, the command signal must be *negative* to energize solenoid *B* and *positive* to energize solenoid *A*. Should these solenoid leads get switched, the spool will move over hard and the valve will open fully.

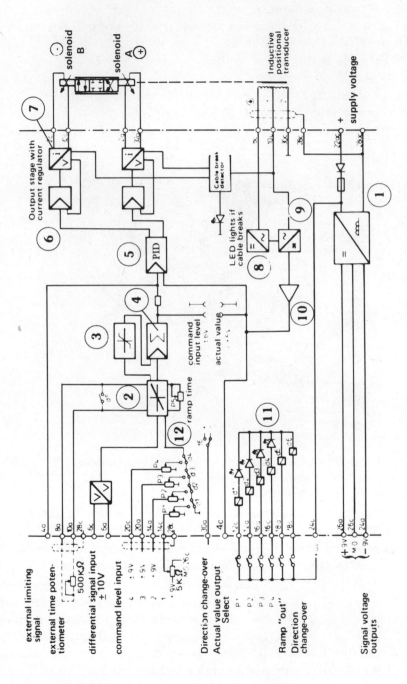

Fig. 6.27 (b) Circuit diagram for amplifier card in Figure 6.27 (a).

The card contains:

1. Voltage regulation and filtering
2. Ramp generator
3. Function generator
4. Summing unit
5. P.I.D. regulator
6. Current regulator

7. Power amplifier
8. Oscillator
9. Demodulator
10. Matching amplifier
11. Relays with LEDs
12. Presets

ELECTRONIC AMPLIFIER FOR ONE STROKE
CONTROLLED SOLENOID PLUS ONE FORCE CONTROLLED SOLENOID

Electronic amplifier one stroke controlled solenoid plus one force controlled solenoid, Figure 6.28 (a), is used to control propor-

Fig. 6.28 (a) Photograph of electronic amplifier card for one stroke controlled solenoid and one force controlled solenoid.

tional, variable volume vane pumps. The amplifier is unique in that it incorporates one circuit to control a *stroke* controlled solenoid and another circuit to control a *force* solenoid, Figure 6.28 (b). Relating this to the proportional vane pump, the top half of the amplifier provides a signal to the main orifice which includes the stroke-controlled solenoid; the bottom half provides a signal to the relief valve which includes the force solenoid.

Internal Circuitry

The top half of the circuit contains:

1. Proportional, integral, differential (P.I.D.) regulator
2. Current regulator
3. Pulse width generator
4. Power amplifier
5. Oscillator, for feeding the LVDT
6. Demodulator, for generating the feedback signal
7. Matching amplifier, for limiting the stroke of the main orifice.

The bottom half of the circuit contains:

8. Current regulator
9. Pulse width generator
10. Power amplifier

The amplifier also contains voltage regulation and filtering, and a ramp generator. Reference voltage terminals provided from the voltage regulation and smoothing block enables potentiometers to provide the required command signal to operate solenoids A and B.

The ramp generator is separated from both circuits. It has an input and output terminal where it can be wired to either set the ramp times for the main orifice or the ramp times of the relief valve.

Presets

$P3$ is an adjustable preset which sets the maximum current or flow to the main orifice; it is accessible on the front face of the card. Although the orifice still can be adjusted remotely by a potentiometer, it will never exceed the value of $P3$, thus avoiding saturation of the output current.

Presets $P4$ and $P5$ function the same way as $P1$ and $P2$ for the single force controlled solenoid amplifier. $P4$ is a minimum current setting that establishes the minimum pressure setting of the valve or the minimum pressure at which the pump compensates. $P5$ is a maximum current setting and establishes the maximum pressure setting of the valve or the pressure at which the pump compensates. Like the single force controlled solenoid, the minimum pressure setting of the valve is additive to the maximum pressure setting, therefore, $P4$ should always be set first.

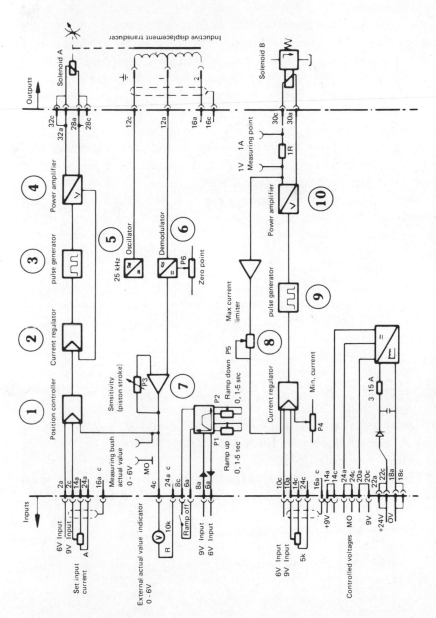

Fig. 6.28 (b) Circuit diagram for amplifier card in Figure 6.28 (a).

Ramp Adjustments

Two ramp adjustments can be used to set separately the speeds at which the orifice opens and closes. The two ramps can also be used to set separately how fast the proportional pressure relief valve will increase and decrease system pressure. The ramp generator can be used for *either* solenoid *A* or *B separately*, but not simultaneously.

Wiring the Ramp Generator

Because the ramp generator is separated from both circuits, wiring must be done externally to include the generator for whatever solenoid it is intended to energize. If the ramp generator is to be used with solenoid *A* (main flow orifice) and remote control is needed for solenoids *A* and *B* by using two potentiometers, the potentiometer that operates the main flow orifice (solenoid *A*) must be connected from the wiper of the potentiometer to 9-volt input terminal of the ramp generator.

The output of the ramp generator is 6 volts, designated by a 6-volt output terminal which must be connected to the 6-volt input terminal 2*a*. The main flow orifice (solenoid *A*) could then be adjusted variably and the ramp generator could be set to control orifice opening and closing times.

The potentiometer that controls solenoid *B* would have its wiper connected directly to 9-volt input terminal 10*a*. Variable adjustments could then be made to the pump compensator. If the ramp generator is to be used for solenoids *A* and *B*, a relay may be added to switch back and forth to operate the ramp generator as desired for a particular solenoid.

There are three sets of test points on the front face of the dual force controlled solenoid Series cards, Figure 6.29. One set reads the

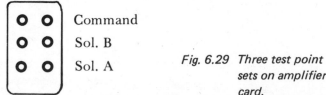

Fig. 6.29 Three test point sets on amplifier card.

command signal; the other two read coil current of solenoids *B* and *A* respectively.

Connecting a voltmeter across the two command test points

lets the operator read the command signal just after the summing amplifier. (Command test points are marked *BU*-1 in the diagram), Figure 6.25 *(b)*. If it appears that an internal relay may not be working properly, the command test points can be used. For example, by energizing each particular internal relay on the card and measuring the voltage across the command test points, the operator can find out which relay on the card is not working.

In the diagram, coil current test points are marked *BU*-2 for solenoid *B* and *BU*-3 for solenoid *A*. Since 1 mA · 1Ω = 1 mV the signal actually being measured is voltage. Therefore, when using the coil current test points, a voltmeter should also be used. If for example the voltmeter reads 0.5 volts, this value is directly proportional to 500 mA. Because a high impedance voltmeter is needed to measure the range of less than 1 volt, it is generally best to use a digital voltmeter.

For one stroke controlled solenoid plus one force controlled solenoid Series cards, two sets of test points are accessible on the front face of the card for measuring the command signal and the feedback signal.

SHIELDING

Because the inputs for all these amplifiers are low level amperage signals ($\pm$ 9V) they are subject to radio frequency interference. The most common way to guard against this problem is to use shielded wire. A shielded wire is a combination of wires with a protective guard around them to help eliminate the intrusion of outside signals.

Shielded wire is usually used when the potentiometer is mounted a few feet from the card, or if the card is mounted in a console and the potentiometer is mounted on top of the console with other electrical devices in the console *i.e.*, relays or anything that can produce magnetic or electrostatic fields. Shielded wire should also be used for wiring the LVDT on stroke controlled solenoids. It is important to remember when using shielded wire, that it should be grounded only at one end. If both ends are grounded, the shield is ineffectual. In fact, it may even *pick up* interferences to compound the problem.

DRY CIRCUIT SWITCH

Anytime a switch is used to provide the $\pm$ 9V-signal, a dry circuit switch should also be used. Since $\pm$ 9V is a low level signal, any dirt that may accumulate around the contacts of a standard switch, may cause them to corrode, rendering the switch useless. Dry circuit switches are designed to work at these low level signals: their contacts are flashed with a gold coating to prevent contamination. Note that shielded wire and dry circuit switches are needed *only* for low level +9 or -9 volt signals. Dry circuit relays are also available.

POWER SUPPLY

When selecting a power supply for an amplifier card the absolute limits should not drop below a minimum of 22 volts or a maximum of 27 volts. The most common type of power supply used to drive the amplifier is a 24-volt DC $\pm$ 10% regulated power supply, Figure 6.30. Although unregulated power supplies may also be used, be

Fig. 6.30 Typical power supply for amplifier card.

aware of the problems that can occur. The more an unregulated power supply is loaded, the greater the voltage drop. Since output voltage is proportional to input voltage, output directly affects the variable being controlled.

Another problem with unregulated power supplies is that if 24 volts are required for a load condition, once the load is removed a variation in voltage occurs which can cause the 24 volts to rise beyond the maximum 32-volt limit, often destroying the card. With the regulated power supply, the output stays at about 24 volts DC even with changes in load. It is, therefore, more desirable to use the regulated power supply. Also remember that if more than one amplifier is used with one power supply, one must carefully check the rating of the power supply to ensure that it is large enough to handle the power capacity of the amplifiers.

CARD HOLDER

Every amplifier card (with one exception) has a 32-pin connector permanently attached to its end. With the connector permanently attached, the card can be plugged directly into a card holder, Figure 6.31, which has two rows of terminals for wiring. Each row contains 16 terminal connections. One row is marked *a*, the other *c*, each row being evenly numbered from 2 to 32. This enables the user to do all the wiring directly to the card holder so that anytime a card must be replaced, this can be done without rewiring.

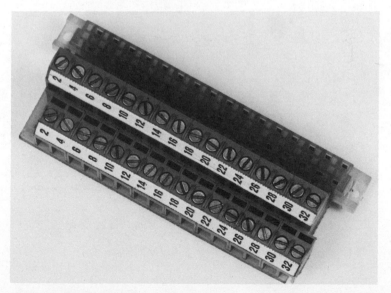

Fig. 6.31 Amplifier card can be plugged directly into card holder equipped with two rows of wiring terminals.

Consequently, if a terminal is marked with *ac*, either *a* or *c* or both can be used for a wiring connection. If, however, two terminals are labeled for one connection, *both* terminals must be used. This is important because each terminal has a limited current carrying ability. To further understand this nomenclature, refer to Figure 6.32.

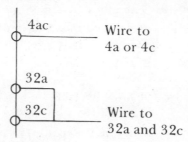

Fig. 6.32 Wiring nomenclature for amplifier cards

If a different card is to be used with a previously wired card holder, make sure that the necessary wiring changes are made, as terminal connections for different amplifier cards may not directly correspond to each other. Not only does the card plug directly into the card holder but guide rails and two fastening screws hold the card firmly in place.

CARD RACK

If multiple amplifier cards are required, rather than mounting each amplifier card to an individual card holder, a card rack, Figure 6.33,

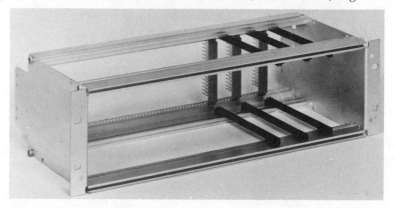

Fig. 6.33 Card rack can hold a number of amplifier cards in one frame.

can hold a number of amplifier cards in one frame. All wiring is then done at the back of the card rack to wire wrapped terminal connections as needed. This method eliminates an excess of terminal connections and keeps the cards in a neat orderly fashion mounting in test consoles. The rack has divisions, so the cards can be spaced equally when required.

AMMETER TEST BOX

For convenient solenoid current testing, an ammeter box, Figure 6.34, with a scale from 0 to 1.5 amps can be used. A special adapter

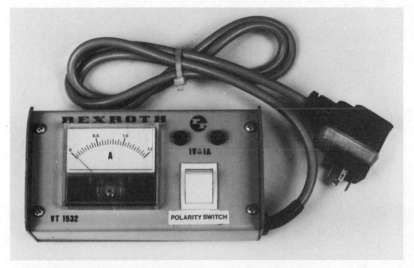

Fig. 6.34 Solenoid ammeter test box has scale from 0 to 1.5 amps.

plug sandwiches it quickly between the solenoid and solenoid plug. Since the plug can be hooked only one way, it also incorporates a polarity button so current can be measured for both solenoids for 4-port proportional directional valves. The ammeter test box is also effective during start-up and trouble shooting to ensure that power is being supplied to the solenoid or solenoids.

REVIEW QUESTIONS

6.1   What two major elements are needed to operate a proportional solenoid?

6.2   What is an electronic amplifier?

6.3   Why is an electronic amplifier needed?

6.4   What does an electronic amplifier do?

6.5   Explain how an electronic amplifier works.

6.6   Is an electronic amplifier the same as an amplifier card. If so, in what way(s)?

6.7   Does voltage regulation affect temperature stability? Explain.

6.8   What is a ramp generator?

6.9   What does a generator do?

6.10 What is a summing amplifier?

6.11 What does a summing amplifier do?

6.12 What are "presets" in amplifier cards?

6.13 Can an amplifier have more than one ramp adjustment? Explain.

6.14 What is meant by *ramp up time* and *ramp down time*? Discuss.

6.15 Define hysteresis.

6.16 Does hysteresis in any way affect valve performance?

6.17 If you think that the answer to question 6.16 is "yes", explain how.

6.18 What causes hysteresis in proportional valves?

6.19 What is meant by valve repeatability?

6.20 How does repeatability relate to a valve's dynamic characteristics?

6.21 Define and discuss dither.

6.22 What is a dither oscillator?

6.23 What is a potentiometer?

6.24 What types of potentiometers are there?

6.25 What are potentiometers used for?

6.26 Is a potentiometer signal linear or rotary? Discuss.

6.27 What does Kirchoff's Law state?

6.28 Is Kirchoff's Law important? If so, why?

6.29 Is an electronic filter similar to a hydraulic filter? Discuss.

6.30 What is a function generator? What does it do?

6.31 What 3 factors cause deadband in proportional valves? Discuss.

6.32 Explain pulse width modulation.

6.33 What is a circuit relay? What does a circuit relay do?

6.34 What does LED stand for?

6.35   What is a limit switch? How are limit switches used?

6.36   Discuss P.I.D. regulators, matching amplifiers and demodulators.

6.37   What does LVDT stand for?

6.38   What do LVDTs do?

6.39   How, where, and why are LVDTs used?

6.40   What is a shielded wire?

6.41   Where and why are shielded wires used?

6.42   What is a dry circuit switch?

6.43   Where and when are dry circuit switches used?

6.44   What is a dry circuit relay?

6.45   What is a card holder? What is it used for?

6.46   What is a card rack? What is it used for?

6.47   When are ammeter test boxes used?